Affect, Space and Animals

In recent years, animals have entered the focus of the social and cultural sciences, resulting in the emergence of the new field of human–animal studies. This book investigates the relationships between humans and animals, paying particular attention to the role of affect, space, and animal subjectivity in diverse human–animal encounters. Written by a team of international scholars, contributions explore current debates concerning animal representation, performativity, and relationality in various texts and practices.

Part I explores how animals are framed as affective, through four case studies that deal with climate change, human–bovine relationships, and human–horse interaction in different contemporary and historical contexts. Part II expands on the issue of relationality and locates encounters within place, mapping the different spaces where human–animal encounters take place. Part III then examines the construction of animal subjectivity and agency to emphasize the way in which animals are conscious and sentient beings capable of experiencing feelings, emotions, and intentions, and active agents whose actions have meaning for the animals themselves.

This book highlights the importance of the ways in which affect enables animal agency and subjectivity to emerge in encounters between humans and animals in different contexts, leading to different configurations. It contributes not only to debates concerning the role of animals in society but also to the epistemological development of the field of human–animal studies.

Jopi Nyman is Head of English at the School of Humanities at the University of Eastern Finland.

Nora Schuurman is Postdoctoral Fellow at the University of Eastern Finland.

Routledge Human–Animal Studies Series
Series edited by Henry Buller
Professor of Geography, University of Exeter, UK

The new *Routledge Human–Animal Studies Series* offers a much-needed forum for original, innovative and cutting edge research and analysis to explore human–animal relations across the social sciences and humanities. Titles within the series are empirically and/or theoretically informed and explore a range of dynamic, captivating, and highly relevant topics, drawing across the humanities and social sciences in an avowedly interdisciplinary perspective. This series will encourage new theoretical perspectives and highlight ground-breaking research that reflects the dynamism and vibrancy of current animal studies. The series is aimed at upper-level undergraduates, researchers, and research students as well as academics and policy-makers across a wide range of social science and humanities disciplines.

Published

Critical Animal Geographies
Politics, intersections, and hierarchies in a multispecies world
Edited by Kathryn Gillespie and Rosemary-Claire Collard

Urban Animals
Crowding in zoocities
By Tora Holmberg

Affect, Space and Animals
Edited by Jopi Nyman and Nora Schuurman

Forthcoming

Animal Housing and Human–Animal Relations
Politics, practices, and infrastructures
Edited by Kristian Bjørkdahl and Tone Druglitrø

Taxidermy and Contemporary Art
By Giovanni Aloi

Affect, Space and Animals

**Edited by Jopi Nyman
and Nora Schuurman**

Routledge
Taylor & Francis Group

LONDON AND NEW YORK

First published 2016
by Routledge
2 Park Square, Milton Park, Abingdon, Oxon OX14 4RN

and by Routledge
711 Third Avenue, New York, NY 10017

First issued in paperback 2017

Routledge is an imprint of the Taylor & Francis Group, an informa business

British Library Cataloguing in Publication Data
A catalogue record for this book is available from the British Library

Library of Congress Cataloging in Publication Data
Names: Nyman, Jopi, 1966- editor of compilation. | Schuurman, Nora, editor of compilation.
Title: Affect, space and animals / edited by Jopi Nyman and Nora Schuurman.
Description: Abingdon, Oxon ; New York, NY : Routledge is an imprint of the Taylor & Francis Group, an Informa business, [2016] | Series: Routledge human-animal studies series | Includes index.
Subjects: LCSH: Human-animal relationships. | Human-animal relationships in literature. | Affect (Psychology) | Affect (Psychology) in literature.
Classification: LCC QL85 .A44 2016 | DDC 591.5--dc23
LC record available at http://lccn.loc.gov/2015021636

ISBN 13: 978-1-138-30834-3 (pbk)
ISBN 13: 978-1-138-92094-1 (hbk)

Typeset in Times
by Saxon Graphics Ltd, Derby

Contents

PART IV
Methodological afterword 179

Figures

Contributors

Lynda Birke is Visiting Professor at the Universities of Chester and Glyndwr, in the United Kingdom. She is a biologist who has worked extensively in human–animal studies. Her most recent research has focused on our relationship with horses, and she has published several articles on cultural change in equestrian culture, as well as on horse–human relationships. This chapter is based on a research project with Jo Hockenhull, and on a keynote presentation to the 'Affective Animals' symposium in Joensuu, Finland, 2013.

Sarah Dean graduated with a bachelor's degree in Anthropology and Spanish from the University of South Dakota. As an undergraduate student, she participated in research on human–animal relationships. She is currently opening an online tutoring business in Las Vegas, Nevada, where she lives with her family.

Karen Dalke is currently Lecturer of Anthropology at the University of Wisconsin-Green Bay in Green Bay, Wisconsin, USA. A primary focus of her research has been the symbolic meaning of the mustang in the American west as the landscape has evolved. Since legislation in 1971, the struggles over protecting the mustang on public lands have illuminated political, economic, and environmental issues between humans and animals. She has been published and has presented papers on this issue in Australia, Finland, Greece, Netherlands, and the United States.

Dona Lee Davis received her PhD in Anthropology from the University of North Carolina at Chapel Hill. She is currently Professor of Anthropology at the University of South Dakota. Her areas of research and publication include maritime, psychological, and medical anthropology as well as gender studies. She has conducted research in Canada, Norway, and the United States. Her most recent book is *Twins Talk: What Twins Tell Us about Person, Self and Society* (Ohio University Press, 2015).

Alex Franklin is Senior Research Fellow at the Centre for Agroecology, Water and Resilience (CAWR), Coventry University and a co-editor of the *Journal of Environmental Policy and Planning*. Her research interests include community resource use, shared practice, human–animal relationships, and skills, knowledge, and learning for sustainability. Based in multi-disciplinary sustainability research centres since 2005, Alex has considerable experience in researching and publishing on environmental practice, both in the UK and internationally. She specializes in qualitative research methods, with much of her work supported by a Participatory Action Research informed approach to study design. Within CAWR Alex has lead responsibility for researching relationships between water and community resilience, with a particular focus on practices and systems of community self-organization.

Sissy Helff is an Anglicist with a broad range of interests in Anglophone world literature, postcolonial and transcultural studies, visual culture, history, and politics. She has published widely including the monograph *Unreliable Truths: Transcultural Homeworlds in Indian Women's Fiction of the Diaspora* (Rodopi, 2013) and the co-edited collections *Global Photographies: History, Memory, Archives* (transcript, 2015), *Films, Graphic Novels and Visuals: Developing Multiliteracies in Foreign Language Education* (LIT, 2013), *Facing the East in the West: Images of Eastern Europe in British Literature, Film and Culture* (Rodopi, 2010), and *Transcultural English Studies: Theories, Fictions, Realities* (Rodopi, 2008).

Jo Hockenhull is a researcher in the Animal Welfare and Behaviour Group at the University of Bristol School of Veterinary Sciences. She has been involved in research on the welfare of UK leisure horses and working equines abroad, as well as on the welfare of farm animals, with particular focus on health status and decisions regarding treatment.

Graham Huggan is Chair of Commonwealth and Postcolonial Literatures in the School of English at the University of Leeds, United Kingdom. He is the author of thirteen books, among them *Nature's Saviours: Celebrity Conservationists in the Television Age* (Routledge/Earthscan 2013), *Postcolonial Ecocriticism: Literature, Animals, Environment* (Routledge 2010, co-authored with Helen Tiffin), and *Australian Literature: Postcolonialism, Racism, Transnationalism* (Oxford University Press, 2007). Most of his recent work is situated at the interface of postcolonial and environmental studies, and he is the co-director of the University of Leeds's new environmental humanities initiative. He is the editor of *The Oxford Handbook of Postcolonial Studies* (Oxford University Press, 2013). He currently leads two big EU-funded collaborative projects, one on Arctic tourism with partners from the UK, Denmark, Iceland, and Norway, and the other in environmental humanities, with partners from Germany, Sweden, and the UK.

Taija Kaarlenkaski received her PhD in Folklore Studies from the University of Eastern Finland, Joensuu campus, in 2012. In her doctoral dissertation, she investigated the construction of human–cow relationships in the material gathered by a public writing competition. In 2013, the University of Eastern Finland granted her the Young Researcher Award for her dissertation. She is currently working on her postdoctoral research concerning the changes in Finnish cattle tending and human–cow relationships from the late nineteenth century to the present day. Her research interests include posthumanist theories and their application in folklore materials, gendered human–animal relations, and emotional aspects of the relationships between cattle tenders and farm animals.

Tua Korhonen, PhD, is Docent of Greek Literature at the University of Helsinki, Finland. She is currently working on the research project Human and Animal in Ancient Greece: Empathy and Encounter in Classical Literature.

Teuvo Laitila is Senior Lecturer in Orthodox Church History at the University of Eastern Finland. He holds a PhD in cultural anthropology from the University of Helsinki and his research interests are the religious history of Karelia and the Balkans. His publications in English include *The Finnish Guard in the Balkans: Heroism, Loyalty and Finnishness in the Russo-Turkish War of 1877–1878 as Recollected in the Memoirs of Finnish Guardsmen* (The Finnish Academy of Sciences and Letters, 2003) and *Nationalism and Orthodoxy: Two Thematic Studies on National Ideologies and Their Interaction with the Church* (with Jyrki Loima; Renvall Institute, 2004).

Karoliina Lummaa is a postdoctoral researcher at the Department of Finnish Literature at the University of Turku, Finland. Her current research project Avian Poetics focuses on the question of nonhuman poetic agency by analysing bird-like formal and thematic features of poetry, e.g., the rhythmic and phonetic features of songs, visual elements resembling movement, and descriptions of avian life and environments. Lummaa's publications include her doctoral thesis *Poliittinen siivekäs: Lintujen konkreettisuus suomalaisessa 1970-luvun ympäristörunoudessa* [*Bird politics: the concreteness of birds in Finnish 1970s environmental poetry*] (University of Jyväskylä, 2010) on Finnish environmental poetry; she has also co-edited several Finnish anthologies devoted to multi-disciplinary environmental research, poetry criticism, and posthumanism.

Anita Maurstad is Professor of Cultural Science at Tromsø University Museum, UiT, The Arctic University of Norway. She received her doctorate from the Norwegian College of Fishery Science in Tromsø. Areas of research and publication include horse–human relationships, materiality, expertise, and museology, as well as small scale fishing and resource management. She has been a rider for 25 years, currently keeping three Icelandic gaited horses.

Jopi Nyman is Professor and Head of English at the School of Humanities at the University of Eastern Finland in Joensuu, Finland. He is the author and editor of seventeen books in the fields of Anglophone Literary and Cultural Studies, including the monographs *Men Alone: Masculinity, Individualism, and Hard-Boiled Fiction* (Rodopi, 1997), *Under English Eyes: Constructions of Englishness in Early Twentieth-Century British Fiction* (Rodopi, 2000), *Postcolonial Animal Tale from Kipling to Coetzee* (Atlantic, 2003), and *Home, Identity, and Mobility in Contemporary Diasporic Fiction* (Rodopi, 2009). His current research interests focus on human–animal studies, transcultural literatures, and border narratives.

Maria Olaussen is Professor of English at the University of Gothenburg in Sweden and Extraordinary Professor at Stellenbosch University in South Africa. Her teaching and research interests focus on African literature, Gender and Postcolonial Studies, with particular emphasis on Southern African literature. This chapter is part of the research project Narrating the Animal Subject: Concurrences as Narrative Strategy funded by the Bank of Sweden Tercentenary Foundation. Her latest book is the edited collection *Africa Writing Europe* (Rodopi, 2009) that deals with representations of Europe in African literature. She has also co-edited a special issue of the journal *Kunapipi* (2012) on African Intellectual Archives as well as a special issue of *English Studies in Africa* (2013) on the works of Abdulrazak Gurnah. She is a member of the research centre on Concurrences in Colonial and Postcolonial Spaces at Linnaeus University.

Nora Schuurman is a Kone Foundation Postdoctoral Research Fellow and Docent at the University of Eastern Finland, Joensuu campus. Her research interests include conceptions of animals, animal-related practices, discourses of animal welfare, and the role of the horse in Western society. Her current research focuses on the affective and performative dimensions of human–horse relationships, in the contexts of leisure horse keeping, equine trade, and horse training, and the practices and conceptions concerning animal death. Her recent research on horses has been published in refereed international journals such as *Humanimalia* (2012), *Society and Animals* (2013), *Anthrozoös* (2014), *Sociologia Ruralis* (2014), *Emotion, Space and Society* (2014), and *Environment and Planning D: Society and Space* (2015).

Jouni Teittinen completed his MA degree (University of Turku, Finland, 2012) in Comparative Literature on romantic posthumanism in the poetics of Charles Olson. His current doctoral thesis concerns the construction of humanity and animality in the post-apocalyptic novel. He is especially interested in how the human as a species impinges on humanity as a cultural and experiential phenomenon, and how this tension relates to views on animality and animals.

Harry Wels is Associate Professor at the Department of Organizational Sciences of VU University Amsterdam and Publication Manager at the African Studies Centre, University of Leiden, both in the Netherlands, and Extraordinary Professor at the University of the Western Cape in South Africa. Trained as an anthropologist, focusing his research on nature conservation issues in South and southern Africa, he increasingly tries to find ways to include animals and animal agency into his analyses. Among his publications is 'Whispering Empathy: Transdisciplinary Reflections on Research Methodology', in *What Makes Us Moral? On the Capacities and Conditions for Being Moral*, ed. B. Musschenga and A. Van Harskamp (Springer, 2013). His current research focuses on human–white lion interaction in the Timbavati area in South Africa, interpreted in the theoretical context of a 'zoopolis'.

Acknowledgments

This book is the result of the research project Companion Animals and the Affective Turn: Reconstructing the Human–Horse Relationship (CONIMAL; project 14875) funded by the Academy of Finland (2011–15) and carried out at the University of Eastern Finland. As the editors of the volume, we should like to express our gratitude to the Academy of Finland for their funding. We should also like to thank our fellow researchers, especially Dr Outi Ratamäki, and all project collaborators for their contribution.

The permission to publish and translate the poetry of Eero Lyyvuo has been granted by Reine Qvist, Tapani Lyyvuo, and Timo Lyyvuo, as well as by the publishing house WSOY in Helsinki, Finland. A section included in Chapter 14 by Karen Dalke and Harry Wels has been published previously in an essay by Harry Wels, 'Whispering empathy: Transdisciplinary reflections on research methodology' included in B. Musschenga and A. Van Harskamp, (eds) *What makes us moral: On the capacities and conditions for being moral*. Dordrecht, Netherlands: Springer, 151–65. We wish to thank Springer Publishers for the opportunity to reuse the section.

Acknowledgement.

1 Introduction

Jopi Nyman and Nora Schuurman

Animals and the spacing of affect

Contemporary life is full of activities involving encounters with diverse animals in diverse places, including everyday contact with rabbits and hedgehogs in gardens and parks, whale watching in the Pacific Northwest and the Azores, sea turtle conservation holidays in Greece, and serious involvement in sheepdog trials and horseracing. Animals attract and move human beings, both on screen and in real life. As a case in point, the website of the National Zoo and Aquarium (2015a) in Canberra, Australia, offers its potential visitors an option to book special visits behind the scenes during which they can '[g]et up close and very personal with some of the worlds [sic] most amazing creatures', including lions, monkeys, and giraffes. What is conspicuous in the description of the zoo tours is that they are invariably described in affective language that appeals to the emotions of the potential visitors. The monkey encounter priced at 100 AUD acquaints the visitor with these 'gorgeous little bundles of cute [sic]' (National Zoo and Aquarium 2015b), whereas the lions and tigers are referred to in a more sublime lexicon as 'magnificent' animals and sources of real 'experience' (National Zoo and Aquarium 2015c and 2015d). Although these visits are commercialized and marketed, and the animals are not in their original habitat, the 'memorable encounter[s]' (National Zoo and Aquarium 2015e) provided by the tours reveal a strong interest in encountering animals in a way that affects the human participant in some personally gratifying way. What appeals to us is the cuteness of the creatures encountered at first hand, personally, through our senses: animals are to be touched and pampered, and their presence affects the visitor in various ways.

Regardless of whether the encounters are material, mediated, or textual, the nature of the encounter, its location, and the way in which it is to be experienced are issues that increasingly intrigue the academic world. In recent years, animals have entered the focus of the social and cultural sciences, resulting in the emergence of the new field of human–animal studies, where the relationships between humans and animals in different contexts are explored. This new wave of scholarly interest in animals has been described as the *animal turn* (Ritvo 2007) and it has led to the publication of novel and often cross-disciplinary understandings of how humans encounter and relate to animals. What is characteristic of the animal turn is that, rather than understanding the animal as playing a passive role in human–animal relations, it places the animal at the centre and perceives it as a subject and an agent contributing to the encounters observed and studied.

Relationships between humans and animals, especially those living within human society, may be scrutinized from the viewpoint of affect, a topical field of discussion in the social and cultural sciences, where its emergence has been characterized as the *affective turn* (Koivunen 2010). The term affect refers to the experience of being affected by the other both bodily and emotionally, an aspect of the human–animal encounter that cannot always be comprehensively understood through language (Ahmed 2010). As such, affect is epitomized in the practices of companion animal keeping, where animals, by entering into individual relationships with humans, serve the emotional needs characterizing life in late modern societies (Franklin 1999: 57). The concept of affect is, however, considered to be more inclusive than that of emotion, when human encounters and communication with non-humans are subjected to investigation (McCormack 2006), and when the reactions are expressed in the context of animal companionship, aesthetics, and suffering, for instance. Emotions, then, can be understood as more cultural and social, and therefore cognitive, although the difference between the two concepts is not straightforward or understood as such. An illustrative definition has been provided by Thien (2005: 451) who understands affects as the *how* of emotions or as 'emotion in motion'. Affects are thus not received in a vacuum but registered in relation to personal experiences and emotions, as Brennan (2004: 7) suggests.

In a human–animal relationship, affect and emotion are not solely the property of humans but the animal is also affected by the human (Nosworthy 2013). What this means is that both humans and animals are able to learn to become affected by the other and change their behaviour accordingly. Such a relational approach to human–animal encounters emphasizes the subjective experiences and actions of the animals in contributing to the production of the relationships in everyday interactions and practices (Birke *et al.* 2004; Haraway 2008). In other words, relationships between humans and individual animals can be understood as co-produced by both human and non-human actors in specific contexts. These contexts are historical and cultural, as well as spatial. Recent work has indeed paid attention to the fact that the location of the encounter, the space where it takes place, influences the ways in which animals are understood and appreciated, and Buller (2014) has suggested that encounters with animals can be understood as spatially situated. Such an approach has recently produced empirical research with a specific focus on the situatedness of human–animal relations, thus adopting the proposition put forward by Philo and Wilbert (2000: 5) that 'the spaces and places involved make a difference to the very constitution of the relations in play'. In general, 'animal spaces' have been defined as either conceptual spaces for animals, addressing the conceptual boundaries and distances between humans and 'other' animals, or as physical spaces, ranging from cities to wilderness and including spaces that are (i) inhabited by animals only, (ii) shared by both humans and animals, and (iii) purely human spaces inaccessible to animals (Philo 1995; Philo and Wilbert 2000: 7–11).

However, as Buller (2014: 310) contends, both the conceptual and material spaces of human–animal relations are not always pre-structured by normative human orderings or otherings. Rather, human–animal relationships are affected by the spaces in which they are performed, often carrying different meanings for human and animal, for example in terms of control and freedom (Schuurman and Franklin 2015). As a result, the visibility of the animal in spaces that have been formerly understood as uniformly human transforms them, and makes them shared or even hybrid spaces where new spatial meanings are provided by both humans and animals. As Philo and Wilbert (2000: 5) suggest, 'animals destabilize, transgress or even resist our human orderings, including spatial ones'. As Nyman (2012) has suggested, the presence of Christian, an African lion bought from Harrods, in metropolitan London, as documented in the memoir *A Lion Called Christian* (2009 [1971]), by Anthony Bourke and John Rendall, constructs new and hybrid spaces where human–animal relations and the conventional roles of the participants are newly reconstituted as a result of a transforming sense of relationality.

Recent research has addressed relationality in detail, and many studies have placed their focus on actual human–animal encounters and individual animals in order to emphasize their subjectivity and agency and, in particular, to foreground the affective agency of animals (Bear 2011; Warkentin and Watson 2014). Interesting examples of such work include Cheryl Nosworthy's (2013) study of disabled horse-riders, which conceptualizes affects as both the emergence of 'flows' before they are captured as conscious emotions and also the capacities that different bodies have for making connections with other bodies; Christopher Bear's (2011) exploration of the ways in which an octopus living in captivity as an individual animal becomes involved in affective relationships with humans in the controlled space of an aquarium, while Emma Power's (2008) work on the constitution of more-than-human families in and through the home, emphasizes the agency of dogs in the process. As the studies reveal, affect can be approached in human–animal relationships by exploring diverse individual encounters, and hence, in this volume, the term is understood more widely, rather than being based on the view of a particular theoretical school. The contributors to this volume approach the notion of affect in various ways and pay close attention to the context of their study.

This volume contains interdisciplinary studies by both humanists and social scientists that examine diverse relationships between humans and animals and pay particular attention to the role of affect, space, and animal subjectivity. Through critical studies of both cultural texts and practices from film and advertising to shared everyday life with animals and comparative religion, the collection shows how the human–animal relationship is configured in diverse contexts and discourses. What is foregrounded is the role of the non-human as a sentient and responsive being, as well as an understanding that the animal is an agent contributing to the construction of the relationship. Through case studies addressing the human–animal relationship, animal-related practices, and animal representations, this volume focuses on the ways in which human–animal encounters are situated, experienced, narrated, and mediated in the contexts of affect and space.

The volume includes three parts and a methodological afterword. The first part discusses the framing of animals as affective, the second part turns its attention to the role of relationality in spatially organized human–animal encounters, and the final part explores various modes of constructing animals and their subjectivity. Several of the essays in the volume take the human–horse relation as their object of study. The transformation in the role of the horse can be understood as exemplifying a considerable number of issues pertinent to contemporary human–animal studies. Human–horse relationships include emotional as well as instrumental values, and equine activities involve encounters with individual animals and the challenge of embodied communication with the animal. Horses are considered as friends whose wellbeing is cared for, and individual human–horse relationships are based on sharing experiences with the animal in specific spaces devoted to horse keeping. As a consequence, the study of human–horse encounters in many cases serves the interests of the wider field of human–animal studies.

Being with animals: affectivity

The chapters in the first part of the volume explore the role of affect in human–animal relations through four case studies that deal with climate change, human–bovine relationships, and human–horse interaction in different contemporary and historical contexts. The readings emphasize the affective bind between humans and animals and the emotional power of the animal image.

In the first chapter, Graham Huggan examines the role of the polar bear, an animal with visible emotional appeal, as an affective animal in the context of current media representations of global warming. Showing how emotionally laden and affective such discourses can be, Huggan foregrounds the symbolism inherent in the ever-popular image of the stranded polar bear. Drawing on the recent confluence of posthumanist 'affective' and 'animal' turns, he understands 'affect' as the capacity to engage meaningfully with others, especially – though not exclusively – through emotional registers. For Huggan, such an understanding of the concept clarifies the various functions of the polar bear in cultural representations of climate change, ranging from advertising to the work of nature documentarists such as Richard Attenborough and Gordon Buchanan.

Taija Kaarlenkaski's ethnological study of human–animal relations in late nineteenth-century rural Finland examines the importance of the cow for the period's smallholders. Drawing on an analysis of ethnographic descriptions concerning cattle tending, folklore, cattle tending guidebooks, and contemporary newspaper articles dealing with humans and cattle, Kaarlenkaski discusses the role played by emotions in transforming relations between humans and cattle amidst modernity. Kaarlenkaski argues that human–animal relationships were ambivalent during the modernization of rural Finland but nevertheless reveal emotional attachments between cattle tenders and their animals. While the physical living conditions of the animals were often poor, and cattle were kept for their products people were attached to them in various situations and in different ways, as indicated in the period's responses to slaughtering and feeding practices and an emerging concern for animal welfare.

The other two chapters address human–horse encounters from two different perspectives. Nora Schuurman and Alex Franklin discuss everyday practices which support the responsible ownership of horses in leisure contexts, including stables and other spaces of human–horse encounters. They explore the possibilities and limitations of interpreting these practices and experiences as they construct the pursuit of serious leisure in the sense proposed by Robert A. Stebbins in his Serious Leisure Perspective (SLP) framework. The chapter pays particular attention to the ways in which the pursuit of horse keeping as serious leisure can benefit the affective relationship between human and horse while also taking into account the horse itself as a sentient being and a subjective actor and agent. Schuurman and Franklin, however, argue that the SLP framework fails to acknowledge relationships of care as forms of serious leisure, and they suggest a revision of the concept with a greater focus on the affective relationship between the animal and its owner, and their actual ways of being together.

In the fourth chapter, Tua Korhonen examines the relationship between humans (and gods) and horses in Homer's warrior society and shows the high value accorded the non-human animal in ancient Greece. Korhonen's analytical rereading of the Homer's *Iliad* reveals that Homer's slightly humanized horses are fully recognizable, flesh-and-blood horses, whose fate in the Trojan War seems to be equated in many ways with that of humans. The warriors address their chariot horses by name; they urge, rebuke, and encourage them; they sometimes even take their sensitivity into consideration; and they also take care of them. Korhonen remarks, however, that the culmination of the human–horse bond in the *Iliad* – Achilleus' relationship with his divine horses – displays more affection on the horses' part than on that of the humans, thus clearly highlighting the agency of horses.

Mapping human–animal spaces: relationality

The second part of the book continues and expands on the issue of relationality but locates the encounters in place and space. It also contributes to an understanding of encounters between humans and animals as products of their practical actions in particular settings – in other words, enactments of human–animal space.

The first chapter in this part of the book, by Jopi Nyman, provides a re-reading of animals, their space, and performativity in Anna Sewell's 1877 novel *Black Beauty*, a classic text promoting animal welfare through popular culture. The chapter demonstrates that the novel's shocking images of animal abuse emphasize affect of a kind typical of the sentimental novel and that this strategy is a form of renegotiating the human–animal relationship amidst Victorian modernity. Reading the novel through the notion of compassion, as discussed by theorists such as Lauren Berlant and Martha Nussbaum, Nyman argues that the novel represents a claim for a new and more equal community where humans and animals may co-exist and where conventional hierarchies and divisions transform.

The chapter suggests that as a part of its critique of discourses promoting human mastery, the novel reveals hybridized spaces of animal performances that reconstruct conventional human–animal relations and aim to restore agency to the non-human. Since humans and horses may perform together and form hybrid and intertwined identities within such spaces, *Black Beauty* seeks to replace conventional understandings of the separateness of humans and animals with an understanding of their togetherness.

The subsequent chapter is concerned with the sentience and awareness of the animal. In her analysis of the novel *The Dog* (1998), by the Swedish writer Kerstin Ekman, Maria Olaussen focuses on issues of memory, time, and place. The chapter follows the theories of the philosophers Matthew Calarco and Giorgio Agamben in identifying literary language as an alternative to scientific and anthropocentric discourses on animals and argues that the tropes and techniques used by Ekman allow for the expression of a specific *Umwelt* of the dog in the sense proposed by Jacob von Uexküll. Ekman's novel explores the interconnectedness of place and time in a story revolving around the agency of the dog as it focuses on its physical needs and sensations in relation to evolving and returning stimuli. The novel also gestures towards the possibilities and limitations of the traditional prose narrative to adequately represent time and place through the awareness of an animal. As a result, what emerges is a dog's mapping of its own animal space.

In the following chapter, Sissy Helff examines contemporary cinematic representations of human–horse relations in three selected feature films set in the late nineteenth and early twentieth centuries: Joe Johnston's *Hidalgo* (2004), Steven Spielberg's *War Horse* (2011), and Belá Tarr's *The Turin Horse* (2010). By approaching the relationship between humans and horses and the underlying politics of transcultural affect in a comparative framework, the chapter draws on individual negotiations of entangled modernities, memory, mourning, melancholia, trauma, and guilt. While the films represent different cinematic styles, Helff sees their foregrounding of affect as a way of expressing modernity's structures of feeling and of unearthing transnational memories of human–horse relationships. In her view, the space of equine representation in the films under study is a transnational and transcultural cinematic space that allows for both political and emotional viewer response.

In the final chapter of the second part of the book, Anita Maurstad, Dona Lee Davis, and Sarah Dean present a multispecies ethnography addressing the situatedness of human–horse relationships where both the material and the mental play a role in the making of the relationship, the human-horse dyad, in movement. Focusing on the equestrian sports of Icelandic gaited horse riding, dressage, and endurance riding, they analyse human and horse engagements as experientially situated in three different ways: situations of being *on* the horse, being *with* it, and being together *in* different terrains and sports. What they suggest is that shared spaces and practices produce a sense of horse-rider togetherness, emphasized in their interviewees' narratives of human–animal interaction as embodied and shared action in a particular environment.

From objects to subjects: exploring animal subjectivity

The third part of the book examines the construction of animal subjectivity and agency through four case studies. The concept of subjectivity as applied in these studies emphasizes the ways in which animals are conscious and sentient beings capable of experiencing feelings, emotions, and intentions, and active agents whose actions have meaning for the animals themselves.

In the first chapter of this part of the book, based on interviews and narratives with equestrians, Lynda Birke and Jo Hockenhull address ways in which horse owners locate their understanding and explanation of horse behaviour in the context of affective human–animal relationships. While the working relationship between horse and human involves learning to work together, this relationship also configures and permits less predictable actions within it, i.e., a kind of breaking through acceptable bounds on the part of the horse. Addressing the reports that the interviewees narrate of their everyday encounters with animals, where they are *moved* in various ways, the chapter shows how the emerging horse stories reveal mutual affect, companionship, and also tolerance of the more behavioural challenges, and how varied interpretations of the horse's personality contribute to the enactment of the relationship. In so doing, it shows that partial connection and separation help to produce different biographies of each partner and the relationship.

In the following chapter, Teuvo Laitila examines the role of Buddhist jātaka tales as means of teaching morally acceptable ways of relating to animal others in classical Buddhist culture. Examining 'The Tale of the Tigress' in particular, Laitila reads the story in the context of human–animal communication and shows the possibility of understanding non-human ways of experiencing the world. The chapter suggests that the tale, which tells about an encounter between Buddha and a hungry tigress, can be read as a version of the theory of mind: humans have a common (biological and existential) basis with other animals, and emotions and reason tell them when the others are suffering. Companionability, Laitila argues, can be based on such understanding, on a belief that the relations that make life less painful are general to everyone. Companionability can thus be seen as a way of building relations aimed at treating the other as a moral agent to the extent of offering one's life to alleviate the other's suffering.

The discussion is followed by Jouni Teittinen's analysis of anthropomorphism and animal suffering through a reading of the American writer James Agee's short story 'A Mother's Tale', a story allegedly told by a bovine mother to its calf. What Teittinen suggests is that Agee's anthropomorphic narrative telling of the horrors of abattoirs foregrounds the power of fiction and allegory in addressing both human and animal suffering rather than merely that of one or the other. Teittinen highlights the ways in which bodily pain and vulnerability become mediated in society, both culturally and technologically, and suggests a reconfiguration of the concept of anthropomorphism in order to appreciate the ways in which representations of animal experience can be performative and have consequences for how we negotiate the limits of description. Teittinen argues that anthropomorphism is not only an epistemic challenge but also a matter of deeply affective ingroup/outgroup control, and that this demarcation is fundamentally tied to the recognition of suffering.

The final chapter by Karoliina Lummaa discusses the poetic, auditive, and linguistic strategies applied in the work of Eero Lyyvuo, a Finnish poet and creator of poetic texts that utilize bird song in providing biographical narratives of the bird imagined in the poem, as forms of naturalcultural poetics. For Lummaa, such bird poetry manages to reconfigure the human and the non-human by paying attention to the non-human's sounds and signs, its experiences and affects, which shows that avian poetry is more than mere animal representation. The affects evoked by Lyyvuo's poetry and naturalcultural poetics more generally are, Lummaa argues, twofold. They are able to create almost ecstatic experiences of sharing and of becoming-animal, and at the same time, they generate discouraging feelings of not understanding, of difference and distance. In sum, the studies in this part show how encounters with animals are intersubjective and open up the possibility of non-human subjectivity and interspecies communication.

Towards new methodologies

The three parts of the volume are followed by an afterword by Karen Dalke and Harry Wels in which they seek to reflect on the applicability of conventional ethnographic and anthropological methods to the study human–animal relations. In discussing the animal turn and its consequences, Dalke and Wels emphasize the importance of empathy as a way of realizing the philosophical concept of becoming as used – in different ways – by authors such as Deleuze and Guattari and Haraway. In an attempt to fulfil the specific requirements of human–animal studies, Dalke and Wels draw on methodological insight generated by scholars in the field of disability studies (e.g., Grandin, Goode, and Sacks) whose studies have surpassed the problem set by the importance of language in conventional research. In the view of Dalke and Wels, the study of trans-species communication with anthropological methods – methods without words – needs more developed non-anthropocentric methods that would be capable of accounting for the experience and perspective of the non-human.

A common feeling among scholars within the field of human–animal studies is that, although the study of animals has become more popular and more respectable, it nevertheless remains marginal. This position is, however, not altogether disadvantageous, as it provides an opportunity to challenge settled assumptions about animals held by both scholars and society at large (Ritvo 2007: 122). Through the case studies and the methodological discussion that it presents, this volume aims to contribute a response to this challenge and to revise the epistemological development of the field of human–animal studies.

References

Ahmed, S., 2010. Creating disturbance: feminism, happiness and affective differences. *In*: M. Liljeström and S. Paasonen, eds. *Working with affect in feminist readings*. London: Routledge, 31–44.

Bear, C., 2011. Being Angelica? Exploring individual animal geographies. *Area*, 43 (3), 297–304.

Birke, L., Bryld, M., and Lykke, N., 2004. Animal performances: an exploration of intersections between feminist science studies and studies of human/animal relationships. *Feminist Theory*, 5 (2), 167–83.

Brennan, T., 2004. *The transmission of affect*. London: Cornell University Press.

Buller, H., 2014. Animal geographies I. *Progress in Human Geography*, 38 (2), 308–18.

Franklin, A., 1999. *Animals and modern cultures: a sociology of human–animal relations in modern society*. london: sage.

Haraway, D., 2008. *When species meet*. Minneapolis, US: University of Minnesota Press.

Koivunen, A., 2010. An affective turn? Reimagining the subject of feminist theory. *In*: M. Liljeström and S. Paasonen, eds. *Working with affect in feminist readings*. London: Routledge, 8–28.

McCormack, D., 2006. For the love of pipes and cables: a response to Deborah Thien. *Area*, 38 (3), 330–2.

National Zoo and Aquarium, 2015a. Animal encounters [online]. National Zoo and Aquarium, Canberra, Australia. Available from: www.nationalzoo.com.au/new_animal_encounters.htm [Accessed 25 February 2015].

National Zoo and Aquarium, 2015b. Book now – meet a monkey – cotton-top tamarind encounter [online]. National Zoo and Aquarium, Canberra, Australia. Available from: https://nazoo.bookingboss.com/booknow.cfm?e=NAZOO1692 [Accessed 26 February 2015].

National Zoo and Aquarium, 2015c. Book now – tiger encounter [online]. National Zoo and Aquarium, Canberra, Australia. Available from: https://nazoo.bookingboss.com/booknow.cfm?e=NAZOO1911 [Accessed 26 February 2015].

National Zoo and Aquarium, 2015d. Book now – white lion encounter [online]. National Zoo and Aquarium, Canberra, Australia. Available from: https://nazoo.bookingboss.com/booknow.cfm?e=NAZOO1588 [Accessed 26 February 2015].

National Zoo and Aquarium, 2015e. Book now – red panda encounter [online]. National Zoo and Aquarium, Canberra, Australia. Available from: https://nazoo.bookingboss.com/booknow.cfm?e=NAZOO1538 [Accessed 26 February 2015].

Nosworthy, C., 2013. *A geography of horse-riding: the spacing of affect, emotion and (dis)ability identity through horse–human encounters*. Newcastle-upon-Tyne, UK: Cambridge Scholars Publishing.

Nyman, J., 2012. From Harrods to Africa: the travels of a lion called Christian. *Society and Animals*, 20 (3), 294–310.

Philo, C., 1995. Animals, geography and the city: notes on inclusions and exclusions. *Environment and Planning D: Society and Space*, 13, 655–81.

Philo, C. and Wilbert, C., 2000. Introduction. *In:* C. Philo and C. Wilbert, eds. *Animal spaces, beastly places: new geographies of human–animal relations*. London: Routledge, 1–36.

Power, E., 2008. Furry families: making a human–dog family through home. *Social and Cultural Geography*, 9, 535–55.

Ritvo, H., 2007. On the animal turn. *Daedalus*, 136 (4), 118–22.

Schuurman, N. and Franklin, A., 2015. Performing expertise in human–animal relationships: performative instability and the role of counterperformance. *Environment and Planning D: Society and Space*, 33 (1), 20–34.

Thien, D., 2005. After or beyond feeling? A consideration of affect and emotion in geography. *Area*, 37 (4), 450–6.

Warkentin, T. and Watson, G.P.L., 2014. Guest editors' introduction. *Society and Animals*, 22 (1), 1–7.

PART I
Being with animals
Affect

2 Never-ending stories, ending narratives

Polar bears, climate change populism, and the recent history of British nature documentary film

Graham Huggan

The polar bear (*Ursus maritimus*) must be one of the most effective animals on Earth. The largest of all bears living today, it is also one of the fastest and most powerful. The top of its food chain, it is a highly efficient predator. Feeding almost exclusively on seals from the Arctic sea-ice platform that provides its primary hunting grounds as well as its main habitat, the polar bear is the epitome of the silent killer, sneaking up on its prey with great skill before striking with fearsome speed and force, enough to flip a 150-pound ringed seal up out of the water and, in the process, crush its skull along with most of the other bones in its body (Feazel 1990: 2–3; see also Bieder 2005: 43). The polar bear is an impressive example of an animal adapted almost perfectly to its environment. Though a land mammal, it swims exceptionally well and is capable of covering vast distances in the water (Lopez 1986). A thick layer of fat, particularly around the rump, helps keep it warm and also aids flotation during swimming, while further fat in its diet (seals) provides the water it needs to process body wastes (Bieder 2005: 44–5). Its paws are significantly larger than those of other bears, 'making efficient snowshoes on land […] and paddles in water' (Bieder 2005: 45), while small protuberances on the pads give further traction on slippery snow and ice. Unsurprisingly, polar bears are treated with the greatest respect by local Inuit, who justifiably fear it – for the polar bear, the world's only animal that will deliberately stalk humans, also has the capacity to strike suddenly and devastatingly, home truths obviously lost on the small but significant number of wildlife tourists in, e.g., northern Canada and Svalbard, who pay the price of their curiosity with their lives (Lemelin and Wiersma 2007).

The polar bear is also one the world's most *affective* animals. Known for some time now as the 'poster child for climate change' (Owen and Swaisgood 2008: 143), it continues to be used in numerous environmentalist campaigns and to feature across a wide range of different media in different countries (as testified in Greenpeace's (2015) recent poster 'Save the Arctic'). The routine use of polar bears for commercial and political ends might help account for the fact that the 'cuddly' polar bear comes second only to the even 'cuddlier' orang-utan as the World Wildlife Fund's most popular adopted animal (Barton 2008; see also Manzo 2010: 198). There seems no limit to its emotional appeal, particularly as a victim (Slocum 2004; Manzo 2010). Some of the most memorable Green campaign images of recent times have been those of bears, frequently caught in sigh-inducing mother-cub poses, being photographed 'stranded' on ice floes – no matter whether they are actually adrift or not.[1]

Polar bears *are* victims, to some extent, insofar as they are caught in life-threatening social and environmental conditions that are not of their own choosing. Their habitat – sea ice – is receding across the vast majority of their range, with most forecasts predicting substantial further reductions; and although population decline has not so far been significant, many of the early warning signs are there, while local population losses, allied to significant habitat alteration, can often be precursors of eventual species extinction in the wild (Derocher *et al.* 2004: 160; see also Owen and Swaisgood 2008; Derocher 2010). However, the polar bear is an affective animal in large part because its appeal transcends the actual conditions in which it lives and forages. As numerous commentators have pointed out, its primary function today is that of an icon of vulnerability in a threatened world; as Robert Bieder suggests, following the American nature writer Barry Lopez (1986: 69), the polar bear is a 'creature of Arctic edges' – with the problem today being that *both* the Arctic, possibly even the world in general for which the Arctic stands as an overdetermined environmental symbol, *and* the polar bears are on the edge (Bieder 2005: 46; see also O'Neill and Hulme 2009; O'Neill and Nicholson-Cole 2009; Höijer 2010; Manzo 2010). The polar bear's affectivity (the measure of its affective state) thus owes not just to its readily exploitable emotional appeal, but to the multi-functionality of its status as an icon: as a physical and visual embodiment of 'the Arctic', 'the wilderness', 'wild nature', and – insofar as it is co-opted as an ecological or anthropomorphic cipher for general imperilment – of 'the planet' or 'humanity' itself.

In this chapter, I want to look at the role of polar bears as 'affective animals' in the context of current (global) media representations of global warming – itself a visibly overheated context in which it has become possible to see the ever-popular image of the stranded polar bear as nothing less than 'the symbol for our times' (Vidal 2008 qtd in Manzo 2010: 197; see also Garfield 2007; Hulme 2009). I will be taking 'affect' here in its broad dictionary sense as the capacity to engage meaningfully with others, especially though not exclusively through emotional registers, and while I will not be dwelling on the considerable complexities of contemporary affect theory, it should be clear from the following that the chapter takes inspiration from the recent confluence of posthumanist 'affective' and 'animal' turns (see, for example, Halley and Clough 2007; also Kalof and Montgomery 2011).

I am similarly indebted to recent work in and across a variety of disciplinary fields on the visual and narrative means currently being used to engage the general public on climate change issues. Empirical work in this area suggests that different images and stories do different kinds of work, and that while a particular image or story may be effective in one area – e.g., *cognition*, the general knowledge of climate science and its underlying issues, it may be relatively ineffective in another, e.g., *affect* or *behaviour*, which are defined in the relevant communications studies literature as the various ways in which individuals understand issues through emotional connection, or in which these understandings, whether primarily cognitive or affective, motivate them to take action of whatever kind (Lorenzoni *et al.* 2007; see also O'Neill and Hulme 2009; Manzo 2010).

As a number of recent UK- and US-based studies suggest, mobilizing the public in both of these countries to act on climate change is itself difficult enough given its vast scale, the uncertainties that continue to surround it, and the fact that it is commonly perceived as affecting 'other' communities in 'other' timeframes, especially the future – all of which makes it hard to connect with at either individual or collective levels in any tangible way (O'Neill and Hulme 2009: 402; see also O'Neill and Nicholson-Cole 2009). In this context, 'anchoring' climate change in particular icons – usefully defined by O'Neill and Hulme (2009: 401) as 'tangible entities [...] bound up with how the viewer relates to [them] through their individual cultural values, sense of place and world view' – can prove to be an effective mechanism in facilitating public engagement with climate change issues, although the effectiveness of these icons (as will hopefully be seen) also largely depends on the stories that are told around them, which operate in turn within larger discursive and ideological frames (O'Neill and Nicholson-Cole 2009; Höijer 2010).

Ending narratives and the 'Aurora effect'

One frame that has been dominant to date is the apocalyptic or 'ending narrative'. The connections between global warming and discourses of environmental apocalypse are well documented (see, for example, Killingsworth and Palmer 1996; also Buell 2010), although, as the theologian Stefan Skrimshire (2010: 220; emphasis added) shrewdly notes, 'apocalypse' may often feature as a temporally *indeterminate* marker rather than, as is more commonly assumed to be the case, an image of 'future *finality*' encapsulated in the end-oriented environmentalist rhetoric of 'tipping points' and 'points of no return'. For Skrimshire,

> [T]he concept or image of future apocalyptic finality [...] functions [today] as something of a smokescreen. Indeed, inasmuch as it encourages the calculation and prediction of prescribed timeframes for viable action, it might also be an illusory consolation, by suggesting that (if a particular timeframe is exceeded) the fight might be over. Far more problematic for us, ethically and politically, is the framing of climate change in terms of the transformation of the present; of the revelation of crisis in our midst with no predictable end [...]. What should interest us about an apocalyptic sensibility are the ethical questions that it generates about ways of living in this indefinite period of waiting and surviving (2010: 220–1).

Skrimshire's critique of the use (and abuse) of apocalypse in the service of climate change alarmism has been picked up, in turn, in some of the communications studies literature on global warming, which draws attention to extreme 'fear appeals' that imply that the problem, even if it is acknowledged as being caused by humans, is beyond the capacity of humans to control (O'Neill and Nicholson-Cole 2009: 358). Not surprisingly, the finger often tends to be pointed here at those sensationalist media practices in which 'fear is employed as a communications tool that [aims to] break through the routine of everyday life and catch the viewer's attention' (O'Neill and Nicholson-Cole 2009: 359) – a largely counterproductive strategy in relation to global warming insofar as it produces dissociation from the phenomenon, e.g., the defensive view that truly dangerous climate change lies elsewhere and/or in a speculative future, making it 'an impersonal and distant issue' (O'Neill and Nicholson-Cole 2009: 361); or the more openly denialist view that its dangers have been greatly exaggerated; or the exasperated feeling that global warming, while of definite concern, is distracting attention away from other issues that can be acted upon more efficiently and closer to home (O'Neill and Nicholson-Cole 2009: 362; see also Hulme 2009; Manzo 2010).

The role of polar bears in fear appeals is not difficult to envision. Iconic images of stranded polar bears arguably elicit public sympathy insofar as they are linked to broader narratives of extinction in which humanity itself is cast as acutely vulnerable, while the recognition of 'our' complicity in 'their' plight provides an example of another kind of 'emotional anchoring' to be found in distressing images of (apparent) animal suffering – that of guilt (Höijer 2010: 719, 723–4). Even more distressing are the images that appear in the so-called 'shock ads' that have been a feature of several recent media-centred climate change campaigns, notably the anti-aviation expansion campaigner Plane Stupid. In one notorious ad, a plane drones overhead as the camera tracks across a series of gleaming tower buildings. Suddenly and without warning, a number of full-size polar bears drop spectacularly from the sky, thudding onto the pavement below in scenes that are clearly calculated to be reminiscent of 9/11 and are no less horrifying (Planestupid.com 2011). The aim of the ad is to draw attention to the fact that every short-haul flight emits 400 kg of carbon dioxide, the equivalent in weight of an adult male polar bear. The ad twists conventional images of animal suffering while relying on the primary strategy (multiple identification) that characterizes polar bear iconicity – for these polar bears are both surrogate humans and unmistakably themselves.

The ad is brazenly exploitative on a number of different fronts, and it inevitably attracted a storm of negative publicity. However, while it is an extreme example it is arguably no less manipulative than other ads of its apocalyptic kind, such as a contemporaneous post-9/11 Greenpeace ad in which a hijacked plane is flown, with suitably terrifying results, into Sizewell nuclear power station. Like most other high-profile climate change campaigners, Greenpeace relies on a variety of mediagenic tactics that attach different forms of *objectification* and *emotional anchoring* to iconic images (Höijer 2010; see also Doyle 2007). 'Objectification', in the words of the Swedish media scholar Birgitta Höijer (2010: 79), 'is a kind of materialization of abstract ideas by representing them as concrete phenomena existing in the physical world', while 'emotional anchoring' refers to those 'communicative processes by which a new phenomenon is attached to well-known positive or negative emotions, for example fear or hope' (Höijer 2010: 79).

While – as is clear from the example – Greenpeace is certainly not above using fear appeals in order to provoke an outraged response to the contemporary social and environmental conditions that are associated with global warming, its tactics of emotional anchoring and objectification cover a wide range of different emotions – fear, pity, compassion, hope, guilt – and, in like fashion, these are visually and rhetorically widespread (Doyle 2007). A good example of a relatively recent Greenpeace campaign that combines several of these tactics is Project Aurora, an international multi-platform operation that joins innovative climate change iconography to a standard 'Save the Arctic' remit. At the centre of the campaign is the giant mechanical figure of Aurora, the 'world's largest polar bear', constructed with state-of-the-art post-*War Horse* technology and paraded, with similarly outsize fanfare, through London's crowded streets (see ODN 2013). Aurora is primarily a hopeful image – a symbol of nature's power – though, like other symbolic representations of its kind, it relies on foreknowledge of other, more-or-less cognate but differently 'emotionally anchored' images: the more elegiac figure of War Horse comes immediately to mind here, as do the brave fighting bears of *His Dark Materials*, which, like *War Horse*, is one of the most successful recent crossover (human/animal) West End stage shows of its kind.

A less well-known intertextual reference, perhaps, is to Aurora World Inc.'s range of cute polar teddies, which range from small to large and even come in pink for girl power.[2] Teddies are perhaps the ultimate 'packaged bear' (Bieder 2005: 102): a reminder of the seemingly universal appeal of bears – or at least stuffed bears – as comforting images. As psychologist Paul Horton (qtd in Crisp 1991: 93; see also Waring 1997; Bieder 2005) has remarked, bears are 'ideally situated in psychological space' insofar as they are 'enough like human[s] for [children] to relate to [them]', but different enough to be distinguished from humans. Teddy bear mania, still alive and well and stretching across several different continents, is one of the world's more curious phenomena: the Teddy Bear Museum in London remains well visited,[3] while not so long ago a stuffed bear collector paid £110,000 at a Christie's auction for a hundred-year-old original Steiff bear (Bieder 2005: 127).[4]

The polar teddy – sweet and lovable, with its imploring, vacuous expression – is about as far as it gets from the 'real thing'; numerous intermediate examples exist, though, notably Knut the bear at Berlin Zoo, who has been 'estimated [probably accurately] to be the biggest cash-grossing animal of all time' (Smyth 2007 qtd in Manzo 2010: 197). As Kate Manzo (2010: 197–8; the embedded quotation is from Connolly 2007) suggests, 'Knutmania' dates back to 2006, when 'the hand-reared bear cub at Berlin Zoo was adopted by Sigmar Gabriel, Germany's environment minister, "in return for using his logo in the campaign against global warming"'. Since then, Knut, who died unexpectedly in 2011, inducing a state of almost Diana-like national anguish, has inspired 'everything from cuddly toys and windscreen cleaners to films and books, [and] has become a symbol for the campaign against global warming, [even appearing] on the front of Vanity Fair' (Connolly 2007 qtd in Manzo 2010: 198).

Manzo and others use the example of Knut – the ultimate celebrity animal – to confirm that 'the role of the polar bear in climate change communication has more to do with affect than cognition' (Manzo 2010: 197), although, like most contemporary communications studies scholars, she adds the caveat that affect and cognition, along with the behaviour they potentially motivate, are – if still distinguishable from one another – mutually dependent realms (O'Neill and Hulme 2009; Manzo 2010). The key point here is made by Höijer (2010: 719), who, while recognizing that some preliminary distinctions can be made between affects and emotions, insists that there is no essential difference between emotion and thinking, and that 'emotions and social cognitions are intrinsically interlinked' (see also Smith and Kirby 2001). However, while empirical studies have amply demonstrated these links, they have also shown that emotional appeal is not necessarily enough to produce the kinds of behaviour that might combat climate change; as Manzo (2010: 199) puts it, 'over-use of the same emotive images may continually re-ignite a compassion response in viewers, but they do nothing for climate change cognition', while 'climate change communication [that is] designed to do more than [produce animal adoptions or] elicit charitable donations cannot continue to rely on stock images of suffering and vulnerability', whether these images are 'emotionally anchored' in compassion or more viscerally dependent on the spectacle of shock (Manzo 2010: 199).

These shortcomings are arguably most visible when stock images of suffering and vulnerability are projected onto 'wild nature'. Such nostalgic visions of the natural world are most likely to be produced today through digital media and television. For more than half a century now, 'nature TV', no more natural a form than the nature it seeks to represent, has served broadly conservationist imperatives, although there is considerable disagreement about its effectiveness in doing so, one frequent criticism being that nature documentary, even as it highlights the technical expertise that makes its own discoveries possible, ironically 'reconfirms the disappearance of animals from the everyday world to which they once belonged' (Berger 1980: 14; see also Wilson 1992; Huggan 2013: 9). The iconicity of the polar bear is singularly well equipped for the mixed demands of modern nature television: it is both 'exotic', occupying a distant space, and visually familiar; while it matches the inquisitiveness of the camera with a curiosity of its own. Most of all, though, it is vulnerable – whether officially 'threatened' or not depends as much on political as scientific interpretations (Owen and Swaisgood 2008) – within the wider contexts of regional and planetary fragility. These are affective contexts in which television, now itself considered to be something of a disappearing medium, operates as a powerful vehicle for registering the miraculous re-appearance of critically endangered animals – including some localized versions of the human animal – even as it mourns the passing of 'authentic' ways of life.

Freeze frames

The first example I want to look at here is the seven-part series *Frozen Planet* (Attenborough 2011 in the UK, 2012 in the US), which forms part of the now legendary BBC-Attenborough franchise, the origins of which are usually traced back to the 1979 natural history epic, *Life on Earth* (Huggan 2013; see also Jeffries 2003). While *Frozen Planet*, like *Life on Earth*, is generally seen as the latest of Attenborough's classic 'blue-chip' documentaries,[5] it might be more accurately described – and is described by Attenborough himself – as a natural history film. Natural history film, as I have argued elsewhere, is 'a specific *kind* of nature documentary, marked by the vast expanse of time and space it covers', and sharing 'some of the underpinning principles of natural history, which, as it emerged in eighteenth-century Europe and developed via the revolutionizing discoveries of Darwin, looked as much to provide models for the moral and political order of human society as a classificatory system for the natural world' (Huggan 2013: 29–30; emphasis added; see also Jardine *et al.* 1996: 8). Natural history film, in other words, is a primarily *moral* form, though not necessarily a didactic one, and Attenborough's view throughout his career has been that his films serve a higher truth and that this aim justifies occasional cheating. (*Frozen Planet*, like several other BBC-Attenborough vehicles, attracted controversy for manipulating the facts. In this particular case, film footage of a polar bear giving birth was discovered not to have been shot on site, but to have been taken from a Dutch animal park – a minor scandal quickly papered over by the filmmakers' not entirely convincing insistence that *in situ* filming would have disturbed the bear and that, in any case, the provenance of the footage had previously been announced.)

Natural history film is also an *observational* form, although *Frozen Planet* continues a sequence begun with the earlier multipart series *Planet Earth* and *Blue Planet* of including a reflexive segment in which the audience is shown how the filming was done. This incremental move towards a more reflexive documentary mode (Nichols 1991) is of a piece with what Alexander Wilson has identified as an industry move from 'nature' to 'science' programming – not that the two are mutually exclusive – in which one of the primary objectives is to transmit basic scientific knowledge about the natural world (Wilson 1992: 146). But as Wilson also points out, it belongs to a longer history, synonymous with 'nature TV' itself, in which technology alternately serves as destroyer and saviour, a double function already apparent in early (1960s) 'drug-and-tag' films in the US, the most popular of which was *Omaha's Wild Kingdom*, in which the formula consisted of 'chasing animals around a savannah in a Land Rover long enough to get some action shots', tranquillizing and caging them once captured, and then hauling them off 'to be studied in the laboratory, where if all went well they would reproduce' (1992: 133).

While 'UK-style' natural history film is clearly not of this swashbuckling kind, and while Attenborough among others has been quick to condemn the more exploitative aspects of the 'drug-and-tag' tradition, it arguably shares at least some of its scientific features, e.g., the view that 'human expertise', particularly technological expertise, is 'necessary for the survival of wildlife' (Wilson 1992: 133). *Frozen Planet*, accordingly, includes sequences that demonstrate the scientific value of monitoring bears, one suitably drugged-and-tagged specimen being given an up-close medical check up.[6] Other polar-bear-centred features of the series are equally formulaic: a stalking sequence and other close shaves; bears on the hunt; cubs at play; a birth scene. These scenes, as might be expected, work the emotions, providing good examples of the different sorts of animal-centred 'emotional anchoring' that I have previously discussed (Höijer 2010; see previous sections of this chapter). More specifically, they play to some of the primary emotions involved in mainstream representations of global warming – particularly compassion, which is the standard commodity at work in scenes such as (taken from the seventh and last part of the series) the formulaically entitled 'On Thin Ice'.[7]

Although there are scattered references to climate change elsewhere in the series, 'On Thin Ice' – initially shelved in the US for fear it might be too politically sensitive for an American audience[8] – is the only one of *Frozen Planet*'s seven episodes to look explicitly at the effects of global warming on circumpolar societies and wildlife. This is in keeping with the series' remit, which, in accordance with the logic of so-called 'last chance tourism', is to chart the natural history of a region before it changes for good (Lemelin *et al.* 2011). As its only half tongue-in-cheek designation implies, 'last chance tourism' represents a lucrative niche-market opportunity to sell (supposedly) vanishing destinations to tourists who are explicitly interested in seeing (allegedly) disappearing natural or social history before it is too late. This market, a good portion of which is climate change driven, is relatively new; but its armchair counterpart, nature television, has long since known how to capitalize on the rhetoric of extinction (which runs from confirmed species loss to the speculative possibility of total planetary destruction) in order to sell its compensatory visions of worlds as they used to be before the irreparable damage wrought upon them by humanity – and the irreparable damage wrought upon humanity by itself (Wilson 1992; Huggan 2013).

The logic of 'last chance tourism' works in two ways, each of which is in direct contradiction to the other. One of its stated objectives is to protect people, animals and landscapes that are acknowledged to be – or are at least popularly believed to be – threatened. But the opposite implication is that they are about to disappear and that it is our moral duty to witness what Wilson (1992: 126) felicitously calls their 'slow recession into history'. In this sense, all wildlife movies – as a secondary form of 'last chance tourism' – are 'a record of lost species, a memento of times and places we [ourselves] have once felt close to in the natural world' (Wilson 1992: 127).

My second example, the BBC2 three-part mini-series *The Polar Bear Family & Me* (Buchanan 2013), provides further evidence of the transferential means by which climate change is 'anchored in nostalgia when [it is depicted] as an existential threat towards people's [or, in this case, animals'] traditional ways of life' (Höijer 2010: 726). Very much at the centre of *The Polar Bear Family & Me* – not a film likely to be approved of by the self-effacing Attenborough – is the excitable Scottish wildlife photographer Gordon Buchanan, whose high-energy TV shows are also notoriously high in sentimental value, and whose filming techniques are frequently invasive (as we will shortly see, dangerous up-close encounters with even more dangerous wild animals are Buchanan's forte, though without the madcap commentary or slapstick antics that tend to characterize TV nature in its postmodern adventure mode: see Huggan 2013, esp. Ch. 5 on Steve Irwin).

The central premise of *The Polar Bear Family & Me* is that Buchanan and his crew follow an increasingly stricken family of bears – a mother and her two cubs, although only one cub seems to be left by the end – over three seasons, beginning with their spectacular emergence from their winter maternity den, and continuing with similarly eye-watering footage of the progressively emaciated mother (groaningly named Lyra, after the eponymous protagonist of Philip Pullman's (2011) Arctic-oriented trilogy, *His Dark Materials*) shepherding her newborn cubs through their first, and highly precarious, year of life. Unlike in *Frozen Planet*, there is little information given on the natural history of the polar bear, and what little is said on climate change is either wrong (e.g., the early claim that the causes of global warming are uncertain) or strategically mealy-mouthed. Instead, we get some cute footage of cubs rolling around in the snow,[9] some frenetic land- and seaborne-pursuit scenes in which Buchanan almost tears up at having finally rediscovered Lyra (Lyra herself seems less delighted), and a lot of unnecessary risk-taking, the most obvious instance of this being a scene in which Buchanan, hunkered down in a specially constructed glass cube, entices a hungry bear to come looking for him – which the bear predictably does, with predictable results.[10]

The scene cost the BBC producers of the show a hefty fine for blatantly contravening local (Svalbard) regulations surrounding the safety zone for viewing large predatory animals; but, more than that, it earned Buchanan himself a torrent of (deserved) criticism, leavened somewhat by (misguided) industry praise. Some of the most hard-hitting views were those of noted polar bear biologists Andrew Derocher and Ian Stirling (2013), whose angry rejoinder to the show is worth quoting in full, though I will restrict myself here to the pithy concluding sentence: 'The focus on the filmmaker, the unethical risks taken, and the unnecessary harassment of wild undisturbed polar bears [all] place *The Polar Bear Family & Me* at the bottom of the scale of natural history reporting'.

It is hard to disagree with this. However, my point is not to skewer the show but to suggest some of the broader difficulties involved in invoking the affectivity of a wild animal when its boundless capacity to *engage* is clearly bound up with the narrower commercial imperative to *exploit*. It is certainly true that, as Owen and Swaisgood (2008: 143) suggest, 'while acknowledging the inherent challenges in linking climate change to specific effects on an individual species, it's hard to imagine a better scientific case than has been built for polar bears'. One obvious problem, though, is that the best (most effective) scientific case does not necessarily create the best (most affective) publicity. Another – just as obvious – is that the different kinds of 'emotional anchoring' that are attached to the polar bear as icon, for all their potential to 'open a window on a global crisis' (Garfield 2007 qtd in Manzo 2010: 197), tend to work in such a way as to reinforce the species boundary that artificially separates us from the animals we only think we know.

Affective engagements of this kind, however sincere, are about our own largely imagined emotional connections to animals, and in imagining ourselves to be reaching out to animals, we are mostly – perhaps even only – reaching out for ourselves. But to end on a more positive note, there may be ways of getting past this seemingly never-ending version of the human-centred human–animal story. One way might be to recognize the value of affect as it pertains to the subjective states and emotional lives of animals; and another might be to admit, as an increasing number of animal scientists do, that affect, far from being 'just a kind of speculation or sloppy thinking that people invoke when a truly "scientific" understanding is not available', has a valuable contribution to make to – indeed, is an integral part of – science's explanatory power (Fraser 2009: 114). Perhaps, in this last context, the best move that most of us non-scientists can make is to separate out *a*ffect and *e*ffect, to acknowledge that animals have rich emotional lives of their own that are irreducible to the uses we (humans) might make of them. And perhaps the next best one might be to join the new 'economy of emotions' (Woodward 2004) to the even newer 'animal turn' (Kalof and Montgomery 2011) in order to see in what ways human and animal lives are ethically, politically, and ecologically entangled; and to hope, as the tangle thickens, that polar bears, who are quite possibly 'the most political of all animals' (Owen and Swaisgood 2008: 123), might find other reasons for not being alone.

Notes

1 Probably the most celebrated individual case is a photograph, first taken in 2004 then reproduced in 2007 when it was released by the Canadian Ice Service, in conjunction with an IPCC report on diminishing sea ice. The photograph was widely discussed, not least because it was argued that the bears were not in fact in any trouble; this then escalated, in some cases, to the further challenging of polar bears as being at risk and, ultimately, of global warming itself. For an account of the history of the photograph and its reception, see Garfield 2007; also Manzo 2010. See also Mouland (2007) and Daily Mail Reporter (2010).

2 See various images on the Aurora Polar Bear site on ebay.com (ebay.com 2015).

3 See The Teddy Bear Museum (2011) and Bieder (2005: 128).

4 Steiff bears, named after their turn-of-the-century German maker Margarete Steiff, quickly became hot properties on the international high-end toy market. As Robert Bieder (2005: 126) notes, 'In 1917, 974,000 Steiff bears were produced and were still not enough. Other toy companies were soon turning them out by the thousands'. Steiff bears are now considered to be valuable collector's items.

5 Blue-chip documentaries are characterized by their high production values, their emphasis on the spectacular and large, charismatic animals, and their use of carefully choreographed set pieces and an emotionally manipulative musical score (Huggan 2013: 32). While there is some argument as to whether Attenborough's natural history films should be considered automatically as blue chip, they are often identified as such and certainly share most of the conventions of the genre.

6 See the clip 'Polar Bear Darting' from the 'On Thin Ice' episode of Attenborough's (2011) *Frozen Planet*.

7 See the clip ''Bleak Future' from the 'On Thin Ice' episode of Attenborough's (2011) *Frozen Planet*.

8 After some prevarication – and a few lame excuses about 'scheduling problems' – the episode was eventually shown.

9 See the clip 'Miki and Luka Playing' from the spring episode of Buchanan's (2013) *The Polar Bear Family & Me*.

10 See the clip 'Close Encounters with a Polar Bear' from the spring episode of Buchanan's (2013) *The Polar Bear Family & Me*.

References

Attenborough, D., 2011. *Frozen planet*. London: BBC Films.

Barton, L., 2008. Our soft spot for the serial killer of the Arctic. *The Guardian*, 10 January, G2, p. 3.

Berger, J., 1980. *About looking*. New York: Pantheon.

Bieder, R.E., 2005. *Bear*. London: Reaktion Books.

Buchanan, G., 2013. *The polar bear family & me*. London: BBC Films.

Buell, F., 2010. A short history of environmental apocalypse. *In*: S. Skrimshire, ed. *Future ethics: climate change and the apocalyptic imagination*. London: Continuum, 13–36.

Connolly, K., 2007. Germans go nuts over Knut, as zoo marks bear's birthday. *The Guardian*, 5 December, p. 25.

Crisp, M., 1991. *Teddy bears in advertising art*. Cumberland, US: Hobby House.

Daily Mail Reporter, 2010. Polar bear and its cub drift on shrinking ice 12 miles from land... but is it all it seems? *Mail Online* [Online], 3 March 2010. Available from: www.dailymail.co.uk/news/article-1254862/All-sea--polar-bear-cub-drift-shrinking-ice-12-miles-land-expert-says-survived.html [Accessed 6 March 2015].

Derocher, A., 2010. The prospects for polar bears. *Nature*, 468, 905–6.

Derocher, A. and Stirling, I., 2013. Reality TV hits new low in the High Arctic with BBC's 'The Polar Bear Family & Me' [online]. Available from: www.polarbearsinternational.org/news-room/pbi-blog/reality-tv-hits-new-low-high-arctic-bbc's-'-polar-bear-family-me [Accessed 2 September 2013].

Derocher, A., Lunn, N.J., and Stirling, I., 2004. Polar bears in a warming climate. *Integrative and Comparative Biology*, 44, 163–76.

Doyle, J., 2007. Picturing the clima(c)tic: Greenpeace and the representational politics of climate change communication. *Science as Culture*, 16, 129–50.

ebay.com, 2015. Aurora polar bear [online]. Available from: www.ebay.com/sch/i.html?_nkw=aurora+polar+bear [Accessed 6 March 2015].

Feazel, C.T., 1990. *White bear: encounters with the master of the Arctic ice*. New York: Henry Holt.

Fraser, D., 2009. Animal behaviour, animal welfare and the scientific study of affect. *Applied Animal Behaviour Science*, 118, 108–17.

Garfield, S., 2007. Living on thin ice. *The Observer Magazine*, 4 March, 32–7.

Greenpeace, 2015. Time is running out. Sign now. Save the Arctic [online]. Available from: www.savethearctic.org/en-US/?utm_source=greenpeace.org&utm_medium=web&utm_campaign=fatfooter [Accessed 6 March 2015].

Halley, J. and Clough, P., eds, 2007. *The affective turn: theorizing the social*. Durham, US: Duke University Press.

Höijer, B., 2010. Emotional anchoring and objectification in the media reporting on climate change. *Public Understanding of Science*, 196, 717–31.

Huggan, G., 2013. *Nature's saviours: celebrity conservationists in the television age*. London: Routledge/Earthscan.

Hulme, M., 2009. *Why we disagree about climate change: understanding controversy, inaction and opportunity*. Cambridge, UK: Cambridge University Press.

Jardine, N., Secord, J.E., and Spary, E.C., eds, 1996. *Cultures of natural history*. Cambridge, UK: Cambridge University Press.

Jeffries, M., 2003. BBC natural history versus science paradigms. *Science as Culture*, 12 (4), 527–45.

Kalof, L. and Montgomery, G., eds, 2011. *Making animal meaning: the animal turn*. East Lansing, US: Michigan State University Press.

Killingsworth, M.J. and Palmer, J.S., 1996. Millennial ecology: the apocalyptic narrative from *Silent Spring* to *Global Warming*. *In*: C.G. Herndl and S.C. Brown, eds. *Green culture: environmental rhetoric in contemporary America*. Madison, US: University of Wisconsin Press, 21–45.

Lemelin, H., Dawson, J., and Stewart, E.J., eds, 2011. *Last chance tourism: adapting tourism opportunities in a changing world*. London: Routledge.

Lemelin, R.H. and Wiersma, E.C., 2007. Gazing upon Nanuk, the polar bear: the social and visual dimensions of the wildlife gaze in Churchill, Manitoba. *Polar Geography*, 30 (1/2), 37–53.

Lopez, B., 1986. *Arctic dreams: imagination and desire in a northern landscape*. New York: Vintage.

Lorenzoni, I., Nicholson-Cole, S., and Turnpenny, J., 2007. Barriers perceived to engaging with climate change among the UK public and their policy implications. *Global Environmental Change*, 17, 445–59.

Manzo, K., 2010. Beyond polar bears? Re-envisioning climate change. *Meteorological Applications*, 17, 196–208.

Mouland, B., 2007. Global warming sees polar bear stranded on melting ice. *Mail Online*, [online], 1 February 2007. Available from: www.dailymail.co.uk/news/article-433170/Global-warming-sees-polar-bears-stranded-melting-ice.html [Accessed 6 March 2015].

Nichols, B., 1991. *Representing reality: issues and concepts in documentary*. Bloomington, US: Indiana University Press.

ODN, 2013. Giant polar bear leads Greenpeace protest to save the Arctic [online]. Available from: www.youtube.com/watch?v=_YO0gfa7Fkw [Accessed 6 March 2015].

O'Neill, S.J. and Hulme, M., 2009. An iconic approach for representing climate change. *Global Environmental Change*, 19, 402–10.

O'Neill, S.J. and Nicholson-Cole, S. 2009. 'Fear Won't Do It': promoting positive engagement with climate change through visual and iconic representations. *Science Communication*, 30 (3), 355–79.

Owen, M.A. and Swaisgood, R.R., 2008. On thin ice: climate change and the future of polar bears. *Biodiversity*, 9 (3/4), 143–8.

Planestupid.com, 2011. Polar bears [online]. Available from: www.youtube.com/watch?v=jTND76fnhyM [Accessed 6 March 2015].

Pullman, P., 2011. *His dark materials*. London: Everyman Books.

Skrimshire, S., 2010. Eternal return of apocalypse. *In*: S. Skrimshire, ed. *Future ethics: climate change and apocalyptic imagination*. London: Continuum, 219–41.

Slocum, H., 2004. Polar bears and energy-efficient lightbulbs: strategies to bring climate change home. *Environment and Planning D: Society and Space*, 22, 413–38.

Smith, C.A. and Kirby, L.D., 2001. Affect and cognitive appraisal processes. *In*: J.P. Forgas, ed. *Handbook of affect and social cognitions*. Mahwah, US: L. Erlbaum, 75–92.

Smyth, J., 2007. Special report: global warming. *British Journal of Photography*, 15 August, 29–32.

The Teddy Bear Museum, 2011. The Teddy Bear Museum [online]. Available from: www.theteddybearmuseum.com/ [Accessed 6 March 2015].

Vidal, J., 2008. The big melt. *The Guardian*, 2 January, p. 21.

Waring, P., 1997. *In praise of teddy bears*. London: Souvenir Books.

Wilson, A., 1992. *The culture of nature: North American landscape from Disney to the Exxon Valdez*. Oxford, UK: Blackwell.

Woodward, K., 2004. *Freud and the passions*. London: Routledge.

3   Cattle tending in the 'good old times'

Human–cow relationships in late nineteenth-century and early twentieth-century Finland

Taija Kaarlenkaski

The starting point of the present study is that there are differing views of the relationships between human beings and farm animals in early modern and modernizing societies.[1] My previous study (Kaarlenkaski 2012) indicated that laypersons often criticize modern industrialized animal production, and see intensive production as a threat to intimate human–animal relationships and attending to cows individually. The latter features were associated with the 'olden days', the time before mechanization, a view also reflected in research (see, e.g., Suutala 2008: 26; Donovan 2013; Klemettilä 2013: 17). On the other hand, some researchers have argued that before modernization farm animals were often treated cruelly in peasant societies and only had instrumental value (Thomas 1983: 147–50; Löfgren 1985: 188; Ritvo 1987: 125–7; Frykman and Löfgren 1987: 75, 79, 179–81; Leinonen 2013: 138, 146–54).

These contradictory interpretations call for more detailed scrutiny. The purpose of this chapter is to discuss this issue by investigating human–cow relationships on small-scale farms in Finland in the late nineteenth century and early twentieth century. Were human–animal relationships actually closer and emotionally more significant in the past than they are today, or were cattle tenders indifferent to their animals and their well-being? This issue has not been studied extensively, as Josephine Donovan (2013: 30n3) has pointed out: 'The relationship between peasants and animals has been largely ignored or overlooked in scholarship on the history of human–animal relations. Works on peasants ignore their relationship with animals and works on animals ignore peasant culture'.[2] Furthermore, interpretations of the relationships between peasants and animals have often been based on descriptions written by the elite, not the peasants themselves. According to the bourgeois world view, the lower classes were indifferent to the suffering of animals, whereas the bourgeoisie were kind to animals and endorsed humane values (Löfgren 1985: 189, 208; Kete 2007; Donovan 2013: 31n4).

In this chapter, I will analyse diverse materials such as ethnographic descriptions, folklore, cattle tending guidebooks, and articles in newspapers. The analysis of materials written from different perspectives casts light on the perceptions of uneducated rural people as well as their educators. Regarding folklore, I will concentrate on descriptions of magic rites used in cattle husbandry. These types of materials have hardly been used in human–animal studies. Seen from the point of view of folklore studies, these materials may reveal shared discourses about the conceptualizations of the animals and of the human–animal relationships formed in cattle husbandry.

Agriculture and animal husbandry in modernizing Finland

The second half of the nineteenth century was a period of modernization, industrialization, and national awakening in Finland. For example, elementary schools were founded and new civic organizations and movements such as the temperance movement, youth clubs, and the labour movement were established. Newspapers written in Finnish started to be published, and they were also increasingly read in the countryside. At the beginning of the twentieth century, the industrialization and modernization processes accelerated, and important political reforms were implemented (Stark 2006a: 9–12; Stark 2011: 38–40).[3] It has been argued that in Finland the late nineteenth century and the early twentieth century created a new 'modern subject [who] was expected to be self-directed and organize his inner impulses and desires to socially productive ends and for the rational benefit of society' (Stark 2011: 39).

At the time, there were also significant changes in agriculture. From the 1870s onwards, farming was modernized by shifting production away from grain growing towards milk production, since the former had become unprofitable due to foreign imports and years of crop failure throughout the 1860s. Dairy production thus became an important source of income for Finnish farmers, unlike earlier, when cattle had been kept primarily for producing manure for grain fields and milk was used only for household consumption. Producing milk for dairies demanded, however, that more attention be paid to hygiene and feeding of the animals. Information on better farming methods was distributed by the press and new farming societies and guidebooks were published (Rasila 2004: 497–9). At the same time, the economy of farms was changing from self-sufficiency towards commercial dairy production. However, the process was slow. The change of the main production sector from grain growing to dairy production took from the 1870s until WWI (Vihola 1991: 12, 31–2). It has been suggested that the 1920s and the 1930s were the 'golden age' of rural Finland, as the production of agricultural products was especially profitable. Between WWI and WWII the production of milk increased by 80 per cent (Jutikkala 1982: 201, 219–21).

In the Nordic countries, a relatively strict gendered division of labour was in place in peasant culture. Tasks outside the farm, such as working in the fields or in the forest, were mostly carried out by men, while women took care of the household and the cattle. While this division of labour was strict in theory, in practice there was a degree of flexibility, but only insofar as women's work was concerned: women could participate in men's work, if needed, and indeed they gained prestige for doing so. However, if men did female tasks, especially milking cows, they would be regarded as unmanly (Thorsen 1986; Markkola 1990: 20–1; Israelsson 2005: 241–6). Although milk started to become financially important for the farms in the late nineteenth century, the traditional division of labour persisted, especially on small farms, until the mid-twentieth century (Thorsen 1986: 139, 142; Östman 2004: 70–2).

Reading the emotional human–animal relationship in old written materials

The materials for this study were produced between the 1860s and 1930s. In order to gain a multidimensional view of the human–animal relationships, the materials represent different perspectives: the viewpoints of cattle tenders as well as those of advisors in the field. The most important materials are ethnographic descriptions, written in response to an ethnographic questionnaire which was first published between 1887 and 1893 and re-published in 1910 and 1930. The questionnaire included a detailed list of questions about different areas of rural life and livelihood, including cattle tending (Haltsonen 1947: 235–6, 349n13).[4] The descriptions were written by laypersons, such as teachers and officers, who were interested in collecting folklore and ethnographic information, but some descriptions were submitted by the farm owners themselves.[5] The formulation of the questions had a clear effect on the material that was gained, as it was usual during those times for the respondents to comply strictly with the questions of the questionnaire (see Schrire 2013). Thus, the responses describe cattle tending on a rather general level; the opinions or experiences of the writers are seldom mentioned (see also Leinonen 2013: 59).

I have studied altogether 68 ethnographic descriptions of cattle tending, written by 61 different respondents.[6] Of the respondents 34 were men and 11 women; 16 writers reported only the initial letters of their first names, but these were probably also men. The descriptions were sent from different parts of Finland, including areas of Karelia which belonged to Finland before WWII and now belong to Russia. However, none of the respondents wrote about Lapland. Nearly half of the descriptions were written in the 1890s or earlier.

Folklore, especially the descriptions of magic rites used in cattle husbandry, provides a different kind of view on human–animal relationships. As ethnologist Laura Stark has put it, magic 'referred to *unnatural* mechanisms for making things happen which derived from secret knowledge or arts' (Stark 2006b: 45; emphasis in original). Magic rites, which often included also verbal incantations, were used in situations where the subject or his/her property was somehow threatened or he/ she wanted to cause harm to other people. During the investigated time period, magic rites were still in use or at least remembered and narrated among Finnish rural people (Stark 2006b: 21–2, 29–30, 45). Using magic rites and word magic in cattle tending was common and a part of everyday practices (Kaarlenkaski 2005; Issakainen 2012). Descriptions of magic rites have been published in *Suomen kansan muinaisia taikoja IV: Karjataikoja 1–3* [Ancient magic rites of the Finnish people IV: Magic rites in cattle tending 1–3] (Rantasalo 1933).[7] In the volumes of this series, the magic rites are organized thematically, and I have gathered those accounts that refer to human–animal communication or conceptualizations of cattle.

To acquire information also from the advisors' point of view and concerning the public discussion regarding cattle tending, cattle tending guidebooks and newspaper articles are used as material as well. Newspapers played an important role in spreading new ideas and farming methods (Stark 2011: 44–6). The articles in newspapers were obtained through the digital collections of the National Library.[8] The database includes a thematic article index which enables finding articles concerning cattle tending, and it is possible to search articles by keywords. The cattle tending guidebooks are published books which were directed at small-scale farmers to improve their methods in animal husbandry. I have included seven books published between 1865 and 1923 concentrating on cattle tending in my research material (see also Kaarlenkaski and Piirainen 2014).

The materials are examined using content analysis and theoretically informed close reading. Special attention is paid to emotional descriptions of cattle and human–animal relations. Emotional expressions may be observed in the texts both at the level of content and formal features. I see emotional experiences as both subjective and constructed in social practices (Harding and Pribram 2009: 4, 12; Latvala and Laurén 2013: 255–6). An illustration of this is that there have been different conceptions concerning what kind of emotions people are allowed to have towards animals. Nowadays emotional relationships are usually associated with companion animals, whereas relationships with farm animals are seen as practical and financial (Franklin 1999: 3; Wilkie 2010: 175). However, emotionality and instrumentality do not exclude each other. Relationships with farm animals have been – and still may be – emotional, especially on small-scale farms where the human–animal relations include personal communication between individuals (Wilkie 2010; Schuurman and Leinonen 2012: 60–1; Kaarlenkaski 2012; Kaarlenkaski 2014).

The challenge of the materials used is, however, that they were not originally produced to describe human–animal relationships or perceptions of cattle. Therefore, finding accounts that refer to these issues was the first task in the present study. After locating the most interesting texts I have read them closely and selected the most central themes. I have actively 'read the animal' in the materials, a method which may be paralleled with resistant reading or sensitive reading (see Fetterley 1978; Latvala and Laurén 2013: 255). The aim is to find and bring forward aspects that are not in the foreground in the texts; meanings that the writers have possibly not even intended to include in them (Lakomäki *et al.* 2011: 12). Reading is focused on the emotional level of the texts as well as on the descriptions of animals and their living circumstances.

The practices of cattle tending

During the time period examined in this study, the farms in Finland were small. In 1920, one half of the farms had less than 5 hectares of field, and only 7 per cent had more than 25 hectares (Jutikkala 1982: 210). There are no statistics from the nineteenth or the early twentieth century that would report on how many cows there were per farm on average, because cattle were owned by different social groups:

farm owners, crofters, and dependant cottagers, and these were put in the statistics separately (Soininen 1974: 204). In 1901, half of the families of the farmhands owned cattle but, usually, they had only one cow (Östman 2004: 57). However, there was regional variation. The ethnographic descriptions mention that on large farms in Western Finland there may have been approximately 20 cows. One respondent from the Karelian Isthmus (Koivisto) wrote that middle-sized farms there had six to eight cows, small farms four or five, and even crofters had two or three cows before WW1. In manor farms there may already have been approximately a hundred cows around the mid-nineteenth century (Vihola 1991: 67–8).

The materials show that the living conditions of the cattle were often poor: cowsheds were dark and damp, and there was insufficient fodder. Before the modernization of animal husbandry, cattle were not given very much hay in the winter; instead, they were fed with straw and mash that included, for example, vegetables, chopped hay or straw, nettle, lichen, and some flour mixed with warm water.[9] According to the ethnographic descriptions, it was quite common to use even human urine, horse excrement or human excrement in the mash. Most of the respondents stated that this was done in the past and were horrified by this practice, but in any case, many of them reported it. Although such feeding was criticized in cattle tending guidebooks, they also recommended straw and root vegetables as part of the diet of cattle. Cows that were milked or were expected to calve soon were fed better than young cattle and cows that were not producing milk at the time.

Under these circumstances, it was understandable that the production of milk was low. It has been estimated that in the 1870s Finnish cows produced approximately 700 litres of milk per year (Rasila 2004: 498).[10] Because fodder was inadequate during winter, cows were usually very weak in the spring when they were let out to pasture. In some cases, they had to be helped to walk out of the cowshed. It has been suggested that the reason for such treatment of cattle was not the laziness or indifference of the farmers, but rational thinking that was based on the contemporary crofting system. Cattle were kept mainly for producing manure for the grain fields, and the number of animals was considered more important than their quality, since they were seen as capital and also as status symbols (Szabó 1986: 27, 37, 39).[11] However, in cattle tending guidebooks and advisory newspaper articles it was often pointed out that it was more profitable to keep a smaller number of cows and feed them well than a larger, poorly fed herd.

Although the new methods for cattle tending were distributed in the press and guidebooks, traditional customs, beliefs, and magic rites persisted in many areas of cattle tending. The coexistence and contradiction of old traditions and new influences is visible in the ethnographic descriptions, newspaper articles, and guidebooks. Because the respondents of the questionnaire were literate and therefore belonged to the enlightened part of the rural population, they often despised these customs and beliefs in their texts. Some of them wrote that in their localities magic rites were not practiced anymore and cattle tending was changing towards more rational practices. As Mikkola (2013: 148) and Stark (2006b: 121, 140) have pointed out, although the amateur collectors of folklore were participants in their local cultures, their ability to write and their interest in literary matters made them observers of their own communities, and they often made use of their own 'modern' identity as a rhetorical strategy in their texts. The use of magic rites was also sometimes reported in newspapers at the time. The tone of these reports was usually disapproving and represented magic rites and their users as old-fashioned.

Nevertheless, folklore collections include thousands of incantations and magic rites related to cattle tending, written down in the late nineteenth century and early twentieth century. The most important situations in which magic rites were used are linked to preventing different kinds of threats: diseases, the evil eye and envy of the neighbours, as well as the dangers of the forests where the cows were pastured. On the other hand, magic rites were also used for acquiring good 'cattle luck' and ensuring success in situations such as calving, breeding, and purchasing cows from other farms (Kaarlenkaski 2005). According to Stark (2006b: 39, 67–9), magic beliefs and practices are linked with ecological and social circumstances in which it is difficult to influence one's own health and living conditions. This was certainly the case in rural Finland during the time period examined. Stark (2006b: 66) has argued that 'the continuous threat of poverty, and the unending labour required of the agrarian populace in order to stay alive' ruled the lives of rural people in the nineteenth century.

As pointed out earlier, one important aspect of traditional farming in Finland was the gendered division of work. However, there are not many explicit descriptions of this division in my research materials. It was probably taken for granted and not even worth mentioning. Moreover, because only a small minority of rural women could write or had the time and equipment to write (Leino-Kaukiainen 2007: 429–35), the voices of the actual cattle tenders are difficult to find in the materials. All of the guidebooks are written by men as are most of the newspaper articles and ethnographic descriptions. The voices of women may be heard in the magic rites, but they were also mostly written down by men. On the other hand, the fact that enlightened men were interested in and wrote about cattle tending may be seen as a signal of change in the gendered division of work.

Emotional relationships?

Generally speaking, there are not many explicit descriptions of emotional human–animal relationships in these materials. One reason for this may be found in the nature of the materials. For example, the questions of the ethnographic questionnaire did not include anything that would refer to emotions. On the contrary, the questionnaire encouraged respondents to write the descriptions on a highly general level and report the practices that were common in the locality of the respondent. Therefore, the descriptions are written mostly in the passive voice and there are hardly any personal comments (see also Thorsen 2003: 55). Another reason for the lack of emotional descriptions may be found in the culture of the time. It has been argued that in peasant culture it was not appropriate to show intense emotional reactions (Siikala 1998: 173, 182). On the other hand, many diaries, autobiographies, and short stories written by self-taught rural people of the time include vivid descriptions of emotions (Kauranen 2009: 11–14; Seutu

2013: 316–20). Thus, it was not uncommon for literate rural people to use emotional language in written texts.

Some clues that refer to emotional human–animal relationships may be found in the materials. In the ethnographic descriptions it is mentioned that the cows were 'dear' to the women. Attachment to the cows as individuals was also seen in the fact that they had proper names and some of them also had pet names (see Phillips 1994). One respondent described milking and reported that the women usually sang to the cows while they milked:

> By singing she expressed that she was kind to the cow. The cow seemed to understand the good treatment and was mutually kind to the milker. It licked the milker and gave its milk to her more easily; whereas, as we know, if you were angry with the cow, it did not usually give its milk. So, harshness did not help, but kindness did (SKS KRA. Eero Väkiparta E 45. 1907. Räisälä, Karelian Isthmus).[12]

It is also mentioned that the cows were talked to and patted while milking – practices that continue to be common decades later. Hand milking, especially, may be perceived as cooperation between the milker and the cow (Israelsson 2005: 142–4; Kaarlenkaski 2012: 216–9, 236–8; Kaarlenkaski 2014: 203). When cows came home from the pastures in the forest in the evening, the farmwife went to meet them and gave them a little food, for example leaves mixed with flour and salted water, from her hand. This was seen as a part of the kind treatment of the cows.

Different kinds of references to emotions may be found in the descriptions of slaughtering. At the time, cattle were usually slaughtered at home by a male resident of the house. In several ethnographic descriptions and descriptions of magic rites it is mentioned that while slaughtering no one present should feel sorry for the animal. According to the belief, if someone in the room bemoans the animal, it dies very slowly and suffers (see also Cserhalmi 2004: 24). These beliefs were related to the traditional practices of slaughtering animals, in which they were not stunned first.[13] According to the descriptions, cows and bulls were usually laid on their sides and tied to a sledge, after which their throats were cut and their blood was let. In some texts, the agony of the animal is described quite graphically and also emotionally.

> The blood runs slowly for about ¼ an hour, the animal groans in agony, makes useless efforts to get away, straightens its legs, dies. When there are no signs of life anymore the slaughterer touches the eye of the animal with his finger or with the point of the knife to test if there is still sensation (SKS KRA. Juho Isopere E 43. 1893. Eurajoki, south-western Finland).

The accounts of the practices of slaughtering tend to be rather laconic, but in the descriptions of dying animals emotional expressions are often used to depict the struggle. It is mentioned that the suffering may have taken as much as an hour, or such a long time that the butcher was able to finish a meal while waiting for the animal to die. The respondents of the questionnaire distanced themselves from such slaughtering practices by denouncing them as cruelty to animals and bewailing the barbarism of the local people (see also Dirke 2000: 205–7).

Although it has been argued that stunning large bovines before bleeding has been a long-standing and widespread practice (Vialles 1994: 17; Higgin, Evans, and Miele 2011: 177), this seems not to have been the case in early modern rural Finland (see also Nieminen 2001: 56). Similar slaughtering methods have also been reported in Sweden (Dirke 2000: 189). Few reasons are given for this practice in the materials, but one respondent to the questionnaire states that the blood runs better if the animal does not die very quickly. According to another, '[i]f the slaughtered animal died immediately, before the blood ran to the ground, the meat remained red even after cooking, and that was not liked' (SKS KRA. Pelkonen, Martta E 126. 1935. Salmi, Ladoga Karelia). These statements suggest that the quality of the meat was given precedence in slaughtering (see also Thomas 1983: 93).

During the late nineteenth century the traditional practice of slaughtering without stunning was already vanishing, and the materials include also descriptions of stunning the animal prior to bleeding by hitting it heavily on the forehead with an axe or by shooting. Advice on stunning was given in the guidebooks and newspapers, and slaughtering courses were also arranged. In 1902, the Slaughter Act was decreed, and it demanded the stunning of the animal (Nieminen 2001: 58–9). According to one ethnographic description, 'old people regard the hitting [on the forehead] crueler than their former practice of torturing' (SKS KRA. Tyyskä, J. E 45. 1893. Askola, Uusimaa). The Slaughter Act was also commented on: 'The newest practice has come into use especially because people are afraid of the power of the law and punishment for cruelty to animals' (SKS KRA. Ruusunen, Nikolaus E 48. 1904. Different municipalities in Satakunta).

The case of slaughtering shows that the emotional reactions towards animals were controlled by traditional beliefs in some areas of animal husbandry. The fact that it was forbidden to feel sorry for the animal while slaughtering may be explained by applying the idea of Laura Stark and seeing certain emotions as agencies or forces. According to her interpretation of the narratives on magic rites, the consequences of negative emotions were not believed to be always controllable by the persons involved. She mentions anger, desire, and envy, but maybe bemoaning could also be this kind of emotion (Stark 2006b: 220–1). Yet, it may be asked why it would be necessary to warn against feeling sorry for the animals if people did not feel compassion for them? It may be argued that the endeavour to control empathy towards the slaughtered animals reveals that the situation was difficult for the people present (see also Cserhalmi 2004: 24). Although the descriptions of emotions are not as explicit as in present-day narratives (Kaarlenkaski 2012: 248–52), these accounts show that there were apparently emotional reactions.

Question of animal welfare

One of the most surprising issues in the materials concerns the use of the concept of animal welfare in a cattle tending guidebook published as early as in 1907. In scientific discussion the concept was not introduced until the 1970s and 1980s, and this is when it was first contested (Broom 2011: 124).[14] The guidebook in question was titled *Ohjeita karjataloudessa pienviljelijöille II: Lehmien ruokinta ja hoito* [Instructions on animal husbandry for small-scale farmers II: Feeding and treatment of cows], and the writer, Hannes Nylander, was an agronomist, specialized in animal husbandry. He worked as a teacher at agricultural schools and wrote several textbooks on the topic. He was also a Member of Parliament, and the title of Counsellor in Agriculture was conferred on him in 1933. Thus, he was a significant figure in the development of animal husbandry and cattle breeding in Finland (*The National Biography of Finland* 2014). The book includes a six-page chapter on the 'Welfare of cows'. The chapter begins:

> We can easily understand how the welfare of our cows depends not only on feeding but on all other external circumstances, such as dwelling, light, quality of air, warmth, physical exercise, rest, etc. Therefore, the profitability of animal husbandry undoubtedly demands that we be able to provide favourable circumstances for our cows in all these respects (Nylander 1907: 35).

According to Nylander, the most important parts of animal welfare were light, fresh air, warmth, cleanliness of both the animals and the cowshed, proper treatment of the cloven hoofs, orderliness in cattle tending, and good treatment of the animals. Although contemporary research expresses the aspects of animal welfare with more abstract concepts, such as health, housing conditions, human handling, and meeting the animal's needs and wants (see Broom 2011: 130; Stamp Dawkins 2012: 142–8), there are evident similarities with these early conceptions.

A later textbook, titled *Lypsykarjan hoito* [Treatment of dairy cattle], which Nylander wrote together with Eino Cajander and Ilmari Poijärvi in 1923, also included a chapter titled 'Animal welfare'. Compared to the earlier book, it contained more detailed instructions on building the cowshed and cleaning. It is important to notice that the arguments for animal welfare in these books refer to the productivity of the animals; the improvements in welfare were not for the sake of the animals themselves. Welfare of the cows meant more income for the farmer (see also Gjerløff 2009: 119). However, when emphasizing the significance of physical exercise, Nylander (1907: 40) wrote: 'It is unnatural that we force animals that are created to be free to stand for 8–9 months per year almost without any exercise'. This indicates that the natural behaviour of the animals was taken into account to some extent. Nylander (1907: 40) recommended letting cattle out of the cowshed to stretch their legs even in winter whenever the weather was suitable.

The concept of animal welfare is not mentioned in other cattle tending guidebooks consulted, but welfare is referred to in some newspaper articles related to cattle tending as early as in the 1890s. In the ethnographic descriptions, however, welfare of cattle was not mentioned with these words, although one aspect of it, good relationships between cattle tenders and the animals, was referred to in several accounts. In the foreword of the first guidebook, Nylander states that the instructions are mostly based on his own experiences, but also on foreign literature. According to Anne Katrine Gjerløff (2009), ideas similar to those presented above were found also in Danish agricultural guidebooks and journals at the time. In any case, these early conceptualizations of animal welfare show that such issues were discussed at the grass roots level of education in animal husbandry at a remarkably early stage.

Conclusion

It has been argued that the attitude of cattle tenders towards their animals has always been practical and caring, since their livelihood has been dependent on the well-being of the animals (Ylimaunu 2002: 121). However, views of good care have changed over time. Practices regarded as customary in the late nineteenth century and early twentieth century seem to be cruel from the present-day perspective. Probably the cattle tenders did their best and cared for their animals, but given the material circumstances in those days animal welfare could not be of a high standard. This is also evident in the materials. According to the texts studied, the contemporaries did not think that they were living in 'the good old times' as regards animal husbandry. On the contrary, the poor state of cattle tending was complained about in ethnographic descriptions, newspapers, and guidebooks; better times were ahead if people would apply new cattle tending methods.

The materials indicate that cattle husbandry at the turn of the twentieth century was characterized by the contradiction between traditional ideas and practices and the information on new cattle tending methods distributed by the press, guidebooks, and farming societies. The development of zoology as well as new theories of breeding also affected animal husbandry in Finland (Kete 2007: 21–2; Toivio 2014). The materials constantly show signs of the coexistence of old and new practices. In the cattle tending guidebooks, for instance, traditional methods were criticized and the aim was to civilize rural people through the promotion of more rational practices (Kaarlenkaski and Piirainen 2014: 14–16). Technical improvement of milk handling and new information about proper feeding of cattle increased the amount and quality of milk and enabled the professionalization of cattle husbandry. It was not merely a part of women's household chores anymore but started to develop into an important source of livelihood (Rasila 2004: 498–9; Kaarlenkaski and Piirainen 2014). This encouraged the farmers to pay more attention to the living conditions of the animals. However, the modernization process was not straightforward and people may have put some new methods and ideas into operation while resisting or ignoring others.

The lack of explicit emotions in the analysed texts may arise at least partly from the nature of the materials. Nevertheless, there are clues that refer to emotional attachments between cattle tenders and their animals. 'Good' and 'kind' human–animal relationships were emphasized in ethnographic descriptions, guidebooks, and newspaper articles. Although the methods of slaughtering were not animal-friendly, possible empathy towards the animals was shown – paradoxically – in the discouragement of feeling sorry for them. According to Niklas Cserhalmi, in

Swedish peasant culture farm animals also had intrinsic value, not just instrumental value. His argument is based on the fact that people were accused of and summoned to appear in court for cruelty to animals. Witnesses who reported these incidents gained no benefit for themselves, which suggests that they felt empathy for the animals (Cserhalmi 2004: 211–38).

It thus seems that human–cattle relationships were ambiguous also 100 or 150 years ago. As Cserhalmi has pointed out, the relationships between human beings and farm animals have always been affected by the conflict between empathy on one hand and productivity requirements on the other hand. He has also argued that the interpretations of human–animal relationships in peasant societies have been ruled by two different approaches. One has emphasized qualitative values, i.e., the close and caring relationships, while the other has been focused on quantitative aspects, such as the quality of food and volume of production (Cserhalmi 2004: 15, 28–9, 357–8). It is evident that if one looks at just one of these features, the views formed of the past will differ. The aim of this chapter has been to take both of these aspects into account. The material circumstances of the cattle were often poor on small farms in modernizing Finland, but that was the case for human beings as well. The animals were kept for the production of food and other products, but they also had a value in their own right and people were attached to them, at least in certain situations. This ambiguity seems to run through the entire history of animal husbandry.

Acknowledgment

I would like to thank the Finnish Cultural Foundation and the Foundation for Advancement of Karelian Culture in Finland for the financial support of my post-doctoral research project.

Notes

1 In Finland, the early modern period ended around the mid-nineteenth century, rather late compared to other European countries. The turn of the twentieth century was an age of societal and cultural modernization. It is important to remember, however, that the changes were gradual and took several decades, especially in the most remote parts of rural Finland (Stark 2006a; Stark 2006b: 462n2).
2 Some studies on human–animal relationships in peasant culture have been conducted in Nordic countries: see Thorsen (2003); Cserhalmi (2004); Israelsson (2005).
3 In 1809, Finland was ceded by Sweden and became an autonomous grand duchy of Russia. In 1917, Finland declared independence.
4 The questionnaire was compiled by a society called Muurahaiset (Ants), established by ethnologist Theodor Schvindt in 1886. The aim of the questionnaire was to collect material for an ethnographic dictionary, but it was never published (Haltsonen 1947: 232–8). The list of questions about cattle tending takes up two pages and includes questions such as 'What kind of food and drink was given to different types of cattle in different situations? [...] What names were given to cows and oxen? [...] How were different types of cattle slaughtered? When, where and how was their life taken?' (*Suomalaisen Kirjallisuuden Seuran kysymyssarja kansatieteellisiä kertomuksia varten* 1930: 38). For ethnographic questionnaires on a more general level, see Schrire (2013).

5 For the motives of the amateur folklore collectors and their relationship to the Finnish Literature Society, see Stark (2006b: 121); Mikkola (2013). On the history of folklore collection and archiving in Finland, see Virtanen and DuBois (2000: 20–7).

6 The original descriptions are stored in the Folklore Archives of the Finnish Literature Society in Helsinki. I have used the microfilm copies, stored in the Joensuu Folklore Archives.

7 On the collection and archiving of descriptions of magic rites, see Stark (2006b: 50–5).

8 The National Library has digitized most of the newspapers and periodicals published in Finland between 1771 and 1910.

9 Hay was given to horses, because they were regarded as more valuable than cows. Horses were used also for public appearance and marking social status, not only working. They were associated with men and belonged to the public sphere, while cows were a part of women's domestic world (Frykman and Löfgren 1987: 181–2).

10 In contrast, in 2013, the average milk production per year was more than tenfold: nearly 8,000 litres (*Yearbook of farm statistics* 2014: 135).

11 Niklas Cserhalmi (2004: 110–82) has criticized this interpretation and pointed out that the narratives about cows that had to be carried to the pasture in the spring were related to years of difficult weather conditions and crop failure. It must be admitted, however, that compared to the standards of modern animal husbandry, feeding was generally poor at the time, regardless of whether the animals actually starved or not.

12 SKS KRA stands for the Folklore Archives of the Finnish Literature Society. Next, there is the name of the respondent, followed by the abbreviation E for ethnographical material and the number of the volume in question. To provide some contextual information, I report the year of submitting the text to the archive and the place where it was gathered. All translations of the original Finnish materials are my own.

13 Practices of slaughtering horses were quite different: they were killed by local skinners who were held in contempt. Horse meat was not eaten and even the horse carcass was regarded as unclean, which affected the position of the skinner. Horses were stunned before bleeding by hitting them on the forehead (Schuurman and Leinonen 2012: 69).

14 The origins of the animal welfare movement may be traced to the founding of the University of London Animal Welfare Society in 1926. The society published its first handbook on the welfare of farm animals in 1966 (Haynes 2011: 106–7). On the history and different conceptualizations of animal welfare, see, e.g., Haynes (2011) and Broom (2011).

References

Broom, D.M., 2011. A history of animal welfare science. *Acta Biotheoretica*, 59, 121–37.

Cserhalmi, N., 2004. *Djuromsorg och djurmisshandel 1860–1925: synen på lantbrukets djur och djurplågeri mellan bonde- och industrisamhälle*. Hedemora, Sweden: Gidlunds förlag.

Dirke, K., 2000. *De värnlösas vänner: den svenska djurskyddsrörelsen 1875–1920*. Stockholm: Almqvist & Wiksell International.

Donovan, J., 2013. Provincial life with animals. *Society and Animals*, 21, 17–33.

Fetterley, J., 1978. *The resisting reader: a feminist approach to American fiction*. Bloomington, US: Indiana University Press.

Franklin, A., 1999. *Animals and modern cultures: a sociology of human–animal relations in modernity*. London: Sage.

Frykman, J. and Löfgren, O., 1987. *Culture builders: a historical anthropology of middle-class life*. New Brunswick, US: Rutgers University Press.

Gjerløff, A.K., 2009. Creating the comfortable cow – discourses on animal protection and production in late 19th-century Danish agriculture. *In*: T. Holmberg, ed. *Investigating human/animal relations in science, culture and work.* Uppsala, Sweden: University of Uppsala, 114–21.

Haltsonen, S., 1947. *Theodor Schvindt: kansatieteilijä ja kotiseuduntutkija.* Helsinki: Kirjapaino-osakeyhtiö Sana.

Harding, J. and Pribram, E.D., 2009. Introduction: the case for a cultural emotion studies. *In*: J. Harding and E.D. Pribram, eds. *Emotions: a cultural studies reader.* London: Routledge, 1–23.

Haynes, R.P., 2011. Competing conceptions of animal welfare and their ethical implications for the treatment of non-human animals. *Acta Biotheoretica*, 59, 105–20.

Higgin, M., Evans, A., and Miele, M., 2011. A good kill: socio-technical organizations of farm animal slaughter. *In*: B. Carter and N. Charles, eds. *Humans and other animals: critical perspectives.* Houndmills, UK: Palgrave Macmillan, 173–94.

Israelsson, C., 2005. *Kor och människor: nötkreatursskötsel och besättningsstorlekar på torp och herrgårdar 1850–1914.* Hedemora, Sweden: Gidlunds förlag.

Issakainen, T., 2012. *Tavallista taikuutta: tulkinta suomalaisten taikojen merkityksistä Mikko Koljosen osaamisen valossa.* Turku, Finland: University of Turku.

Jutikkala, E., 1982. Omavaraiseen maatalouteen. *In*: J. Ahvenainen, E. Pihkala, and V. Rasila, eds. *Suomen taloushistoria 2: teollistuva Suomi.* Helsinki: Tammi.

Kaarlenkaski, T., 2005. Karjataikuus – naisten työtä, akkojen puuhastelua. Master's thesis in Folklore Studies. Joensuu, Finland: University of Joensuu, Department of Finnish and Cultural Research.

Kaarlenkaski, T., 2012. *Kertomuksia lehmästä: tutkimus ihmisen ja kotieläimen kulttuurisen suhteen rakentumisesta.* Joensuu, Finland: Suomen Kansantietouden Tutkijain Seura.

Kaarlenkaski, T., 2014. Communicating with the cow: human–animal interaction in written narratives. *In*: K. Tüür and M. Tønnessen, eds. *The semiotics of animal representations.* Amsterdam: Rodopi, 191–216.

Kaarlenkaski, T. and Piirainen, M., 2014. Hyötyä ja hyvinvointia kansalle: 1800–1900-lukujen vaihteen puutarhan- ja karjanhoidon opaskirjat aineistona. *Elore* [Online], 21 (2), 1–22. Available from: www.elore.fi/arkisto/2_14/kaarlenkaski-piirainen.pdf. [Accessed: 15th January 2015].

Kauranen, K., 2009. Menneisyyden muistiinpanojen kirjo. *In*: Kauranen, K., ed. *Työtä ja rakkautta: kansanmiesten päiväkirjoja 1834–1937.* Helsinki: Finnish Literature Society, 6–21.

Kete, K., 2007. Introduction: animals and human empire. *In*: K. Kete, ed. *A cultural history of animals in the age of empire.* Oxford, UK: Berg, 1–24.

Klemettilä, H., 2013. *Federigon haukka ja muita keskiajan eläimiä.* Jyväskylä, Finland: Atena.

Lakomäki, S., Latvala, P., and Laurén, K., 2011. Menetelmien jäljillä. *In*: S. Lakomäki, P. Latvala, and K. Laurén, eds. *Tekstien rajoilla: monitieteisiä näkökulmia kirjoitettuihin aineistoihin.* Helsinki: Finnish Literature Society, 7–27.

Latvala, P. and Laurén, K., 2013. The sensitive interpretation of emotions: methodological perspectives on studying meanings in oral history texts. *In*: Frog, P. Latvala, and H.F. Leslie, eds. *Approaching methodology.* Helsinki: Finnish Academy of Science and Letters, 249–66.

Leino-Kaukiainen, P., 2007. Suomalaisten kirjalliset taidot autonomian kaudella. *Historiallinen aikakauskirja*, 105, 420–38.

Leinonen, R-M., 2013. *Palvelijasta terapeutiksi: ihmisen ja hevosen suhteen muuttuvat kulttuuriset mallit Suomessa.* Oulu, Finland: University of Oulu.

Löfgren, O., 1985. Our friends in nature: class and animal symbolism. *Ethnos*, 50, 184–213.

Markkola, P., 1990. Women in rural society in the 19th and 20th centuries. *In*: M. Manninen and P. Setälä, eds. *Lady with the bow: the story of Finnish women.* Helsinki: Otava, 17–29.

Mikkola, K., 2013. Self-taught collectors of folklore and their challenge to archival authority. *In*: A. Kuismin and M.J. Driscoll, eds. *White field, black seeds: Nordic literary practices in the long nineteenth century*. Helsinki: Finnish Literature Society, 146–57.

The National Biography of Finland, 2014 [Online]. Maanviljelysneuvos Hannes Nylander (1873–1940). Available from: www.kansallisbiografia.fi/talousvaikuttajat/?iid=386. [Accessed: 12 February 2013].

National Library of Finland, Centre of Preservation and Digitization. Newspapers [online]. Available from: http://digi.kansalliskirjasto.fi/sanomalehti/secure/main.html [Accessed: 2 March 2015].

Nieminen, H., 2001. *Sata vuotta eläinten puolesta: kertomus Suomen Eläinsuojeluyhdistyksen toiminnasta 1901–2001*. Helsinki: Suomen eläinsuojeluyhdistys.

Nylander, H., 1907. *Ohjeita karjataloudessa pienviljelijöille II: lypsylehmien ruokinta ja hoito*. Porvoo, Finland: Werner Söderström Osakeyhtiö.

Nylander, H., Cajander, E., and Poijärvi, I., 1923. *Lypsykarjan hoito*. Helsinki: Otava.

Östman, A.-C., 2004. Mekanisoinnin ensimmäinen aalto. *In*: M. Peltonen, ed. *Suomen maatalouden historia II: kasvun ja kriisien aika 1870-luvulta 1950-luvulle*. Helsinki: Finnish Literature Society, 146–57.

Phillips, M.T., 1994. Proper names and the social construction of biography: the negative case of laboratory animals. *Qualitative Sociology*, 17, 119–42.

Rantasalo, A.V., [compiler] 1933. *Suomen kansan muinaisia taikoja IV: karjataikoja 1–3*. Helsinki: Finnish Literature Society.

Rasila, V., 2004. Overview of the history of Finnish agriculture – from prehistory to the 21st century. *In*: P. Markkola, ed. *Suomen maatalouden historia III*. Helsinki: Finnish Literature Society, 490–507.

Ritvo, H., 1987. *The animal estate: the English and other creatures in the Victorian age*. Cambridge, US: Harvard University Press.

Schrire, D., 2013. Ethnographic questionnaires: after method, after questions. *In*: Frog, P. Latvala, and H.F. Leslie, eds. *Approaching methodology*. Helsinki: Finnish Academy of Science and Letters, 201–12.

Schuurman, N. and Leinonen, R.-M., 2012. The death of the horse: transforming conceptions and practices in Finland. *Humanimalia: A Journal of Human/Animal Interface Studies* [online], 4 (1), 59–82. Available from: www.depauw.edu/humanimalia/issue%2007/schuurman-leinonen.html [Accessed 19 February 2015].

Seutu, K., 2013. Novellit: Kirjallistumisen merkit kansankirjoittajien novelleissa. *In*: L. Laitinen and K. Mikkola, eds. *Kynällä kyntäjät: kansan kirjallistuminen 1800-luvun Suomessa*. Helsinki: Finnish Literature Society, 304–32.

Siikala, A.-L., 1998. Oliko savolaisilla tunteita? *In*: J. Pöysä and A.-L. Siikala, eds. *Amor, genus & familia: kirjoituksia kansanperinteestä*. Helsinki: Finnish Literature Society, 165–92.

SKS KRA (Folklore Archives of the Finnish Literature Society). Ethnographical material, volumes E 1, E 42–8, E 51, E 58, E 61, E 64, E 67, E 101, E 103, E 122, E 126–7, E 132–3, E 135, E 137–39, E 144, KRK 221, KT 33.

Soininen, A.M., 1974. *Vanha maataloutemme: maatalous ja maatalousväestö Suomessa perinnäisen maatalouden loppukaudella 1720-luvulta 1870-luvulle*. Helsinki: Suomen Historiallinen Seura.

Stamp Dawkins, M., 2012. *Why animals matter: animal consciousness, animal welfare, and human well-being*. Oxford, UK: Oxford University Press.

Stark, L., 2006a. Johdanto: pitkospuita modernisaation suolle. *In*: L. Stark and S. Tuomaala, eds. *Modernisaatio ja kansan kokemus Suomessa 1860–1960*. Helsinki: Finnish Literature Society, 9–46.

Stark, L., 2006b. *The magical self: body, society and the supernatural in early modern rural Finland.* Helsinki: Academia Scientiarum Fennica.

Stark, L., 2011. *The limits of patriarchy: how female networks of pilfering and gossip sparked the first debates on rural gender rights in the 19th-century Finnish-language press.* Helsinki: Finnish Literature Society.

Suomalaisen Kirjallisuuden Seuran kysymyssarja kansatieteellisiä kertomuksia varten, 1930. Helsinki: Kirjapaino Osakeyhtiö Alfa.

Suutala, M., 2008. *Onnellisia lehmiä ja viisaita ihmisiä. –Elämää saariston luontaistaloudessa ja sen muutoksia.* Vihti, Finland: Karprint oy.

Szabó, M., 1986. Hade djuren det bättre förr? *In*: A. Biörnstad, ed. *Husdjuren och vi.* Stockholm: Nordiska museet, 27–50.

Thomas, K., 1983. *Man and the natural world: changing attitudes in England 1500–1800.* London: Penguin Books.

Thorsen, L.-E., 1986. Work and gender: the sexual division of labour and farmers' attitudes to labour in central Norway, 1920–1980. *Ethnologia Europaea,* 16 (2), 137–48.

Thorsen, L.-E., 2003. Glad i dyr? En diskusjon av dyriske følelser och følende dyr. *Tidsskrift for kulturforskning,* 2 (4), 51–64.

Toivio, H., 2014. Risteytyksistä maatiaisrotuihin: professori Victor Prosch ja kotieläinjalostuksen murros 1800-luvun jälkipuolella. *Lähde: historiatieteellinen aikakauskirja,* 10, 96–122.

Vialles, N., 1994. *Animal to edible.* Cambridge, UK: Cambridge University Press.

Vihola, T., 1991. *Leipäviljasta lypsykarjaan: maatalouden tuotantosuunnan muutos Suomessa 1870-luvulta ensimmäisen maailmansodan vuosiin.* Helsinki: Suomen Historiallinen Seura.

Virtanen, L., and DuBois, T., 2000. *Finnish folklore.* Helsinki: Finnish Literature Society in Association with the University of Washington Press, Seattle, US.

Wilkie, R., 2010. *Livestock/deadstock: working with farm animals from birth to slaughter.* Philadelphia, US: Temple University Press.

Yearbook of farm statistics, 2014. Helsinki: Information Centre of the Ministry of Agriculture and Forestry.

Ylimaunu, J., 2002. Elinkeinot ihmisen ja eläimen suhteen muokkaajana. *In*: H. Ilomäki and O. Lauhakangas, eds. *Eläin ihmisen mielenmaisemassa.* Helsinki: Finnish Literature Society, 115–33.

4 In pursuit of meaningful human–horse relations

Responsible horse ownership in a leisure context

Nora Schuurman and Alex Franklin

The practice of keeping horses is increasingly understood as a leisure activity, but what does it actually mean to think of horses as animals of leisure? In engaging with this question this chapter concentrates on the ownership of horses by non-professionals or 'amateurs'; that is by those individuals who engage in horse ownership solely as a form of leisure practice, to be undertaken entirely in their free time. It is among this category, referred to here as 'leisure horse owners' that by far the greatest rise in horse ownership is currently occurring in the Western world.[1] The human–horse relationship has always been based on individual encounters and communication between the two, and in contemporary equestrian culture, the affective aspect of the relationship is emphasized and the horse is increasingly perceived as a companion. The expectations of horse ownership are focused on the emotional and affective qualities of the human–horse relationship and on forming a partnership with the horse, alongside the practical use of the horse for purposes of sport and competition.

Understanding the construction of any individual human–animal relationship as never 'fixed', always in the becoming, this chapter takes as its focus the construction of a mutually rewarding human–horse relationship, leading to what is often termed as 'responsible horse ownership' by animal welfare organisations and educational institutions. The discussion is guided by the conceptual framework of Robert A. Stebbins's (1992; 2001; 2012) Serious Leisure Perspective (SLP). Stebbins's concept is useful for studying this new wave of leisure horse ownership as, by encouraging reflection on whether or not it can be understood as a 'serious' leisure practice, it supports a review of levels of commitment to the human–horse relationship and equine welfare in contemporary horse keeping culture.

We pay particular attention to the ways in which the pursuit of horse keeping as serious leisure can benefit the affective relationship between human and horse by taking into account the horse itself as a sentient being and a subjective actor and agent. To do this, we seek to conceptualize the practice of keeping horses from the emergent relationalist view point in human–animal studies, turning the focus towards the actual relationship between the human and the animal. We begin by asking whether a human–horse relationship, including the idea of responsible horse ownership, can be understood as 'serious leisure'. Simultaneously, though, we re-evaluate the typology presented by Stebbins in his SLP, with the introduction of the animal other, an element that challenges several of Stebbins's very concrete assumptions of the empirical reality of leisure. This in turn leads us to ask, to what extent might individual horse–human relationships impact upon an individual owner's commitment to the daily practicing of serious leisure? Central to the analysis is the horse, an active agent in influencing the pursuit of a human–horse relationship as serious leisure, as well as the welfare implications of the above-mentioned practices for the horses.

The relationality of responsible horse ownership

The rise of amateur equestrianism and leisure horse keeping in recent decades can be understood as part of the new sensibilities towards animals, leading to increased pet keeping and other animal-related activities (Franklin 1999). Keeping a horse opens up the possibility for affective encounters and individual bonds with an animal, as well as opportunities to participate in the variety of equestrian sports. But, what actually is a 'responsible horse owner'? In contemporary equestrian culture in the West, responsibility in horse ownership most of all refers to questions concerning equine welfare. With animal welfare becoming all the more topical in animal-related activities, discussions about equine welfare are also an inseparable part of the culture of keeping horses. In equestrianism the task of caring for the horse in the best possible way is widely acknowledged, as well as the requirements of special knowledge and personal experience to achieve this.

Within the UK, for example, the official equine welfare code of practice (Defra 2009), published in direct follow-on from the 2006 Animal Welfare Act, advises prospective owners that prior to purchasing a horse, they should first review their 'existing level of experience and whether they have the skills and knowledge to care for a horse properly', with the recommendation that where relevant 'consideration should be given to gaining prior experience with horses via riding stables or through undertaking voluntary work' (Defra 2009: 2). What is perhaps particularly significant about the new wave of horse owners is that for as many who could be characterized as confident and experienced in handling horses, there exist many more who, at the time of purchasing a horse, are not. For these, such is the apparent lure of purchasing their own horse, and 'being' with them on their own terms, that an absence of practical knowledge or knowledgeable practice (Ingold 2000) in the art of good horsemanship is seemingly perceived as no actual barrier.

The extent to which a fullness of experience and skills required for comprehensive horse care can be gained through contact with horses in the context of a riding school (where horse contact is generally limited to the set time of a riding lesson) is questionable (Birke 2008). Such is the nature of horsemanship, that in the absence of regular opportunities for spending extended periods of time in the direct company of horses and in so doing gaining first-hand experience of routine duties of care – to the point also of being able to understand, cater for, and respond to variances in such as behavioural characteristics and dispositions of individual animals – there is a limit to how much of this practice-based knowledge and understanding can actually be developed. This is primarily because in the context of keeping horses, tacit knowledge has been the predominant type of knowledge until the present day (McShane and Tarr 2007: 39; Birke 2008). Knowledge about horses is largely embodied (Brandt 2004) and based on material practices and encountering the animal. It requires personal experience of individual horses acquired over time, and therefore cannot be achieved solely from written sources such as scientific writing and the media.

According to several recent studies (Barad 2003; Birke *et al.* 2004; Irvine 2004; Haraway 2008), knowledge of the animal other, especially tacit, personal, or embodied knowledge, is an essential prerequisite for a successful human–animal relationship. This is also the case in individual human–horse relationships that are largely based on embodied communication between the two (Brandt 2004; Birke 2008). Such knowledge can also be understood as a manifestation of the relationality and affect characteristic in human–animal encounters. As an epistemological starting point, relationality refers to

> a transactional dynamic interrelationship [...] taking place between knower and known and capable of yielding *enough* awareness of the latter by the former to enable negotiation, or better a navigation of what phenomenologists call the 'lifeworld' – a domain or zone of experience shared with other forms of life (Acampora 2001: 74, emphasis original).

Thus, to be in a relationship and to take care of the other, one has to know the other. Knowing the animal and being able to communicate with it involves a sharing of lived experiences with the animal, or co-habitation with the animal to some extent. The spaces of everyday human–animal relations thus become significant as settings for shared everyday life.

A relational approach takes into account animal agency, as animals themselves, by their subjective actions, are seen to participate in the production of the relationships (Barad 2003; Birke *et al.* 2004; Haraway 2008). As Haraway (2008: 36) points out, '[a]ccountability, caring for, being affected, and entering into responsibility are not ethical abstractions; these mundane, prosaic things are the result of having truck with each other'. Here, affect refers to a mutual process where both human and horse are affected by the other in bodily encounters and shared activities, resulting in the development of an emotional relationship (Nosworthy 2013: 59). While the desire to gain proficiency in horsemanship may not necessary act as a core motivator towards horse ownership, ownership is nevertheless the most opportune route to gaining such proficiency and often reveals its importance in the daily routines soon after the purchase. Moreover, core to the domains of equine practice accessible through ownership, is the possibility of developing a deeper, more meaningful, and sustained relationship with an individual horse, than would commonly be possible with those available for hourly riding hire through trekking centres or riding stables. A meaningful human–horse relationship requires a different spatial setting, in other words spaces of affect.

The SLP, animals, and the lack of affect

What are the implications for horses and their human partners of the growing (human) desire for participating more 'seriously' in equine-based leisure practices by means of actual horse ownership? As noted above, the term 'serious' is used here in reference to the SLP framework put forward by Stebbins. Despite the considerable duration of literature on serious leisure, as yet barely any attempt has seemingly been made to apply Stebbins's framework to a study of human–animal relations. In responding to this gap we consider here whether such an application can inform us about keeping animals for leisure and the affective significance and welfare implications of such activities for both animal and human. Is there a possibility for 'serious human–animal leisure' and if so how is such practice shaped and produced by the relationship between human and animal? Before engaging with these questions, though, we begin with a more detailed introduction to the principle of *serious leisure*.

According to Stebbins, the concept of serious leisure can be defined as 'the systematic pursuit of an amateur, hobbyist, or volunteer activity sufficiently substantial, interesting, and fulfilling for the participant to find a (leisure) career there acquiring and expressing a combination of its special skills, knowledge, and experience' (2012: 69).

As such, serious leisure can be understood as the opposite of *casual leisure*, which is 'immediately, intrinsically rewarding, relatively short-lived pleasurable activity requiring little or no special training to enjoy it' (Stebbins 2001: 53). Unlike casual leisure, serious leisure has the capacity to be deeply satisfying and to give a feeling of full existence (Stebbins 2001: 54). Stebbins (2012: 69–71) classifies serious leisure into three separate categories: i) *amateurism* found in art, science, sport, and entertainment, ii) *hobbyist activities* such as collecting, making and tinkering, participating in non-competitive action, hobbyist sports and games, liberal arts hobbies, and iii) *volunteerism*, providing a service to others, either in an informal or formal setting.

In this chapter, our focus is limited to whether it is possible to apply Stebbins's concept to affective, individual human–animal relations. In Stebbins's typology, animals are found in two examples of serious leisure. One of them is in the category of volunteerism, more precisely in the sub-category of 'faunal volunteering'. Defined as an activity for 'those interested in certain kind of animals' (Stebbins 2012: 72), the context or situation where the activity in question might occur is not specified, whether it be feeding backyard birds in the winter or working with an animal rights group. Following Stebbins, however, the category of volunteerism is problematic in the context of individual human–animal relationships due to the fact that it excludes family members from the ones that volunteerism may benefit. As members of the communities of their owners, animals contribute to the emotional, relational, moral, and practical aspects of the activities they are kept for, and companion animals are therefore increasingly understood as family members (Charles and Davies 2011). In many cases this also applies to horses kept for leisure.

The other example, the one closest to horse ownership, is 'horseback riding', classified in the hobbyist section of non-competitive nature hobbies, with the aim of 'appreciation of the outdoors' (Stebbins 2012: 71). Evidently, keeping companion animals has not occurred to Stebbins as demanding special attention in his definition of serious leisure. Furthermore, in the few empirical studies of serious leisure in the context of animal activities, the focus has been on the human social community of, for example, dog sports (Gillespie *et al.* 2002). What has been under scrutiny is the human world of dog owners, but not their relationship with the actual animals. We suggest, therefore, that bringing the practice of keeping horses for leisure into the discussion of serious leisure has a capacity to challenge Stebbins's strict typology by bringing the question of affect into the discussion. Our suggestion stems from the multi-layered nature of the human–horse relationship, consisting of affective encounters, self-interest, and altruistic motivations.

Responsible horse ownership as a career in serious leisure?

Stebbins (1992; 2012: 72–5) gives serious leisure six definitive qualities: a need to persevere despite occasional adversity, the opportunity to follow a career in the endeavour, a significant personal effort based on acquired knowledge, training, or skill, several durable benefits for participants, a unique ethos within the community of participants, and a distinctive identity. In addition, Stebbins discusses the question of motivation in serious leisure at length. In the following sections, we concentrate on the possibility of understanding the development of a deep and meaningful relationship with a horse, including knowledge of the animal, as a career in serious leisure in the sense suggested by Stebbins.

According to Stebbins (2012: 81), a career in serious leisure is 'the typical course, or passage, of a type of amateur, hobbyist, or volunteer that carries the person into and through a leisure role and possibly into and through a work role'.[2] While Stebbins (2012: 82–3) concentrates on classifying the different phases of a leisure career, or the participants in serious leisure according to their dedication and interest in their pursuits, here we are interested in the qualities of the career and the possibilities of developing it into a criterion for serious human–animal leisure in the context of responsible horse ownership.

Similar to volunteering, activities with animals are motivated both by self-interest and altruism, two seemingly opposite but closely intertwined motives (Stebbins 2004: 2). In the context of human–animal relations, these motives are visible in the well-known ambivalence inherent in animal attitudes, namely the controversy between using animals for human benefit and attributing emotional and moral attitudes to them (Franklin 1999). This is also the case in amateur equestrianism and leisure horse keeping as they are often motivated by the success attainable by training the horse and riding it in competitions. However, keeping a horse is not simply about self-interest or personal benefit. Instead, it is necessarily about a more complex intertwining of self-interest and altruism where the responsibility for the wellbeing and ultimately the survival of the animal are in the hands of the owner. The horse aptly epitomizes the phenomenon described by Arluke and Sanders (1996: 131) that seemingly opposite systems of meaning, such as emotionality and instrumentality in human–animal relations, can exist simultaneously in practices involving animals. Spending time with individual animals, working with, and getting to know them tends to involve a process of mutual affect and bring about an emotional bond with the animal, whatever the fate of the animal in that activity may be (Wilkie 2010). The ambivalent status of the horse is also apparent in contemporary discussions on animal welfare. The need to care for one's horse properly, to be a responsible owner, is a recognized part of the performance of contemporary horse ownership (Birke 2008).

It is in the process of becoming a responsible horse owner that we find the opportunity to achieve a serious leisure career, as opposed to a series of casual leisure engagements. To explain this further we draw on the concept of animal capital developed by Leslie Irvine (2004) in her study of relationships between humans and companion animals. According to Irvine (2004: 65), animal capital comes from curiosity 'about how animals communicate, feel and learn, and about what they need to live healthy lives'. Animal capital includes knowledge and skill needed in keeping animals, on topics such as behaviour, health, nutrition, and training, as well as an understanding of their emotions, communication, and cognition. Furthermore, a central aspect of animal capital is a curiosity about the animal and 'knowledge about how to find things out' (Irvine 2004: 66), for example by seeking veterinarians or trainers when in need of support. A 'serious' horse owner would thus not only be a skilful horse handler and rider, but also have an active interest in bonding with the horse and taking care of it.

Irvine (2004: 66) argues that animal capital produces a different kind of relationship with animals as the concept of animal capital is not value-free but refers to 'resources that enable the development of meaningful, non-exploitive companionship with animals'. Irvine (2004: 67) later points out that animal capital is not about knowing the animal in order to exploit it, but instead recognizes the intrinsic value of the animal. From this viewpoint, the attitude or commitment of the horse owner towards the horse makes a difference. It is ultimately the animal that benefits from the efforts of the owner, which again renders the relationship meaningful for both human and horse. Thus, the knowledge, skill, and curiosity about the animal and the relationship, supported by a pursuit of equine welfare, can be understood as responsible horse ownership leading to a career in serious human–animal leisure. Characteristic of such a career is a long-term learning process situated within affective encounters, in other words 'a form of mutual becoming which occurs over time' (Thompson 2011: 232). Following this, the concept of animal capital serves as a usable criterion for serious leisure in horse ownership.

But how can differentiation between serious leisure and casual leisure be achieved in such a rather abstract context? To do this, we follow Irvine's discussion about cases where something goes wrong. The owner's reactions to crises of different kinds in the relationship offer a practical viewpoint into defining seriousness in horse ownership. Shifting our attention towards this more practical focus, in turn, also allows us to address our second research question. That is, to what extent might individual horse–human relationships impact upon an individual owner's commitment to the daily practicing of serious leisure?

The risk of failure within serious human–animal leisure

A motivation for pursuing a serious leisure career is, for Stebbins (2012: 75), a mix of the costs and rewards – personal fulfilment – experienced in the activity. Stebbins concludes that the rewards are ultimately higher than any costs experienced. For Stebbins's SLP to be of more value, however, the framework is first in need of further development such that it supports considerably more attention being given to what may happen, what the consequences may be, when something goes wrong. This is an acknowledged risk in relationships with animals, with possible problems ranging from welfare issues to failed communication and open conflict. Considering the possibility of failure is therefore important in the context of an understanding of responsible horse ownership as pertaining to the pursuit and maintenance of equine welfare. In cases of injury or illness in the horse (or rider), or when the cooperation between human and horse fails to work, the question of motivation may be more complex than Stebbins has suggested, and this is so exactly because of the affective character of the relationship.

In the view of Stebbins (2012: 5), 'Leisure is typically conceived of as a positive mind set, composed of, among other sentiments, pleasant expectations and recollections of activities and situations'. As he also goes on to note, though: 'Of course, it happens at times that expectations turn out to be unrealistic, and we get bored (or perhaps angry, frightened, or embarrassed) with the activity in question, transforming it in our view into something quite other than leisure' (2012: 5–6). Owning a horse involves taking responsibility for the daily care, handling and training of the horse, taking into account its individual needs, temperament and life history, and managing any unexpected challenges in its health or behaviour. If, as discussed above, the step up to horse ownership is preceded only, for example, by occasional contact with a horse in the context of a riding school lesson, then it is certainly foreseeable that the 'pleasant expectations and recollections' may have little bearing on the reality of serious leisure. Moreover, where it turns out that the initial selection of a horse for purchase is not sufficiently compatible with the abilities, resources (especially time), and interests of a particular owner, what then are the implications for the welfare of the horse or wellbeing of the owner? Unfortunately, Stebbins does not go on to discuss what this 'something quite other than leisure' might be, nor for that matter, what happens when a shift occurs from perceiving an activity as leisure to regarding it as non-leisure, the permanency of such a shift, or whether it might again come to be viewed as leisure at some point in the future.

A review of the UK code of practice for equine welfare suggests a failure to adequately cater for the situation whereby, at the point of becoming a horse owner, many individuals possess only very rudimentary knowledge of horse behaviour or the practices of care. This failing is reflective of the lack of attention given within welfare guides in general to the nature of the relationship between human and animal. Instead, there are recommendations throughout the UK equine welfare code for contacting an expert whenever in doubt, seen in this example from the short section on equine behaviour: 'If you are unsure how to best handle your horse, advice should be sought from an experienced horse professional' (Defra 2009: 15).

While injury and illness can without doubt generate feelings of guilt and shame in the owner (Birke *et al.* 2010), the role of veterinarians and veterinary science in supporting the recovery of the horse (and the owner) is significant. Problems in communication with the horse are not so easily understood, defined, and solved. Further, the physical size and strength of horses pose a risk of injury to people interacting with them (Brandt 2004). Therefore, in all activities involving horses, control is in human hands, and power, meaning the domination of horses by humans interacting with them, is an inherent feature of human–horse relations (Thompson 2011). In discussions of serious leisure, Stebbins's theory has been criticized for not recognizing the power relations inherent in the activities studied (Raisborough 1999). It is, thus, reasonable to ask whether power and control exercised over an animal contribute to the enthusiasm felt in interacting with horses. As Haraway (2008: 26) points out, embodied communication with animals is not always easy or harmonious. What happens in cases where efforts at controlling the animal turn into an open conflict with the animal, and the relationship is not rewarding anymore but filled with fear and mistrust?

In the previous section, we defined the concept of animal capital as a criterion for serious human–animal leisure. A closer look at the roots of Irvine's concept reveals a concern for failure in relationships with companion animals. Irvine (2004: 66) argues that in failed relationships, two factors are involved: i) a commitment to the animal as a subjective being and an active participant in the person's life, and ii) the inability or failure to seek and use resources to cope with a problem. In other words, the persistent effort to solve a problem is also a sign of commitment to and responsibility for the animal and the strength of the affective relationship. An owner who has actively invested in knowledge of both the animal itself and of 'how to find things out' is also more likely to succeed in seeking help and solving problems. Considering serious leisure from this viewpoint, this is where serious human–animal leisure really turns serious. The motivation to persist and to struggle through unpleasant challenges in order not to lose what is an essential element in one's life, namely the animal itself, should be taken into account in the definitions of serious leisure.

It should be recognized that leisure is not ethical as such (Blackshaw 2010: 40), thus something belonging to the sphere of serious leisure is not necessarily an act of good will. Blackshaw (2010: 43–4) points out that Stebbins positions serious leisure as something profound and 'good', and its opposite, casual leisure, as superficial and worthless. It should be remembered, however, that serious leisure, while having the opportunity to become ethical, can also turn adverse. This is especially the case in the context of animal care, with considerable controversy in the understandings of what is good for the animal (Birke 2008; Birke *et al.* 2010). Without the curiosity and willingness to find things out for the benefit of the animal, the possibility for responsible horse ownership is considerably limited.

Spaces of serious human–animal leisure

Being a responsible horse owner seemingly requires a commitment to a level of horse care provision which conforms to more than a casual leisure approach. In revising the concept of serious leisure such that it may act as a conceptual guide for exploring the bundle of practices which come together in the form of responsible horse ownership, we have set out (above) the reasons as to why it is more instructive to focus more centrally on the affective relationship between horse and owner; that is their ways of being together. Before drawing this argument to a conclusion, however, it is useful first to spend a moment connecting the preceding review of responsible leisure horse ownership to a consideration of the structural and spatial context in which much of the growth in leisure horse ownership appears to be taking place.

Facilitating the rise in leisure horse owners, including those with only limited existing knowledge or experience of horse care, has been an associated growth in the number of horse livery yards. Reflective, at the same time, of the mass of owners who themselves reside in urban areas, this increase is particularly notable within the peri-urban and rural spaces located at the city's edge. Notable in their role as significant structural supports for opening up horse ownership to a much wider cross-section of the public, is the fact that livery yards make ownership viable not only for those with limited previous experience or knowledge, but also for those with restricted (or irregular) amounts of free time which they are able to commit to spending in the company of their horse (as dependent, for example, on work pressures, or family commitments). In addition to providing stabling and grazing space, many yards also include the option of a range of horse care service packages, which can be tailored to the needs of the owner (and horse), on a monthly, weekly, or even daily basis. Although cost and available time are certainly relevant factors in shaping individual livery arrangements, the preference shown by many owners is, wherever possible (i.e. leisure time permitting), to address the daily care needs of their horse themselves. This suggests that a core aspect of the attraction of horse ownership is the opportunity for being directly involved in the process of daily care provision and therefore, spending extended periods of time in the company of their horse, on a regular basis and in a shared space, approaching the idea of co-habitation with the animal.

Further, there is seemingly a potential for livery yards themselves to support the aspiration for responsible horse ownership and a career in serious leisure by individual horse owners. Apart from the provision of facilities and expertise by the yard owner, access and exposure to professional help from outside is often easier on a livery yard. At the same time, the community of horse owners can share their knowledge and assist in practical tasks, while their horses can also benefit from each other's company, a welfare aspect in itself. The role of community in serious leisure is acknowledged by Stebbins (2012: 73–5), but only in a sense of a social world or subculture with a common goal. Instead, a livery yard can be understood as a place where individual human–horse relationships come together, each with their own goals that are nevertheless similar. More research is needed to find out the extent and ways in which yards are actually able to enhance the pursuits of horse owners for serious leisure and whether it is experienced by the horse owners in this way. As a starting point and as a way of acknowledging the specific networks of human–animal, human–human, and animal–animal relations emerging at a livery yard, a socio-spatial approach would be beneficial, one that addresses the spaces and places of serious human–animal leisure.

In spite of the shortcomings of Stebbins's binary categorisation, his concept of serious leisure has proved its potential as an analytical tool for assessing the passions, values, and meanings associated with pursuits characterized as affective, such as horse ownership. However, we suggest a revising of the concept of serious leisure such that it may act as a conceptual guide for exploring the practices which come together in the form of responsible horse ownership. The idea of animal capital by Irvine (2004) serves to highlight this issue. In empirical studies of serious human–animal leisure, it is instructive to focus more centrally on the emotional and affective relationship between the animal and its owner, and their actual ways of being together.

There are, however, also some notable limits to the usefulness of the SLP for exploring occurrences of responsible horse ownership. Most notably, as currently conceptualized by Stebbins the framework fails to acknowledge relationships of care as forms of serious leisure. The main concern here is the difficulty of fitting the affective human–horse relationship and the horse itself, including the owner's efforts in (or neglect of) caring for the horse's welfare, into the idea of serious leisure. In this sense, the concept reflects an individualist attitude inherent in much of leisure studies. In leisure theory, the central focus has been on benefits for the individual, such as health and wellbeing (Arai and Pedlar 2003). Similarly, the concept of serious leisure emphasizes both individual activity and benefits, including benefits to the individual from social interaction (Gallant *et al.* 2013). Arai and Pedlar (2003: 187) note that the main problem with individualism is the failure to acknowledge the questions of interdependence and obligations in society, and social justice. This also concerns leisure if leisure activities are understood as being principally based on consumption and personal benefit. In discussions of leisure, unencumbered individualism has led to the acceptance of individual benefits, choice, and autonomy – in other words, independence – as the goals of leisure activities. Instead of consumerism, Arai and Pedlar (2003: 190) suggest that attention be directed to focal practices based on shared meanings and cooperation, and their argument, while restricted to humans, can easily be extended to non-humans.

The idea of serious leisure seems well grounded for such notions, but these are not developed within Stebbins's principle. Instead, serious leisure seems to have a rather narrow individualist focus. According to Gallant *et al.* (2013: 323), this is because 'serious leisure developed amid increasing individualism and related societal and leisure trends of commodification and consumption'. In the recent book *Serious Leisure and Individuality*, Cohen-Gewerc and Stebbins (2013: 10) do acknowledge the threat of falling into what they call the 'trap of individualism's abyss'. They do not, however, elaborate on how this can be avoided in the pursuit of freedom and one's unique individuality. Furthermore, despite writing about leisure and consumerism at length, the writers do little more than claim that leisure should be understood merely as a way of consuming goods, with serious leisure concentrating on goods needed in the activities.

In Stebbins's model, any altruistic action within serious leisure is restricted to volunteerism, and this poses some questions concerning the inclusion of animal activities in the concept. A human–horse relationship based on mutual affect, shared meanings, and care for the animal will necessarily include some altruistic action. A strictly individualist mode of keeping horses is in contrast with the idea of knowing the animal and being interested in its life and wellbeing the way Irvine suggests. Altruistic actions towards one's own animal, such as taking care of its welfare, can therefore be understood as challenging the individualism inherent in the very definitions of serious leisure. Thus, not only do we need to revise the concept of serious leisure, but also extend the scrutiny of altruistic action to other forms of leisure than solely volunteerism. Only then would a relationship with an animal be considered serious leisure, with a potential for personal benefits to both human and animal.

To conclude, in order for the concept of serious leisure to be of greater assistance in supporting our understanding of the pursuit of meaningful human–animal relations, we suggest a revision of the current understanding of leisure in leisure studies. The recent transformation in the ways of relating to animals in free time is an indication of multiple values that are also present in leisure activities. In defining 'leisure as activity' Stebbins (2012: 6) writes that 'an activity is a type of pursuit, wherein participants in it mentally or physically (often both) think or do something, motivated by the hope of achieving a desired end'. Instead of a definite goal, however, a human–animal relationship is never fixed, but always in the becoming (Haraway 2008). Coupled with the pursuit of animal welfare and mutual knowing and sharing, serious human–animal leisure challenges the instrumental and individualist idea of leisure in favour of a more affective, communicative, and altruistic understanding of leisure in the Western world.

Notes

1 By ownership we also refer to long-term loan arrangements where someone is responsible for the care and use of a horse owned by someone else.
2 In equestrianism, the development from amateur to professional through personal experience is not uncommon.

References

Acampora, R.R., 2001. Real animals? An inquiry on behalf of relational zoöntology. *Human Ecology Review,* 8 (2), 73–78.
Arai, S. and Pedlar, A., 2003. Moving beyond individualism in leisure theory: a critical analysis of concepts of community and social engagement. *Leisure Studies,* 22 (3), 185–202.
Arluke, A. and Sanders, C.R., 1996. *Regarding animals.* Philadelphia, US: Temple University Press.
Barad, K., 2003. Posthumanist performativity: toward an understanding of how matter comes to matter. *Signs: Journal of Women in Culture and Society,* 28 (3), 801–31.
Birke, L., 2008. Talking about horses: control and freedom in the world of 'natural horsemanship'. *Society and Animals,* 16 (2), 107–26.

Birke, L., Bryld, M., and Lykke, N., 2004. Animal performances: an exploration of intersections between feminist science studies and studies of human/animal relationships. *Feminist Theory*, 5 (2), 167–83.

Birke, L., Hockenhull, J., and Creighton, E., 2010. The horse's tale: narratives of caring for/about horses. *Society and Animals*, 18 (4), 331–47.

Blackshaw, T., 2010. *Leisure*. London: Routledge.

Brandt, K., 2004. A language of their own: an interactionist approach to human–horse communication. *Society and Animals*, 12 (4), 299–315.

Charles, N. and Davies, C.A., 2011. My family and other animals: pets as kin. *In:* B. Carter and N. Charles, eds. *Human and other animals: critical perspectives*. Basingstoke, UK: Palgrave Macmillan, 69–92.

Cohen-Gewerc, E. and Stebbins, R.A., 2013. *Serious leisure and individuality*. Montreal, Canada: McGill-Queen's University Press.

Defra, 2009. *Code of practice for the welfare of horses, ponies, donkeys and their hybrids* [online]. London: Department for Environment, Food and Rural Affairs. Available from: www.gov.uk/government/uploads/system/uploads/attachment_data/file/69389/pb13334-cop-horse-091204.pdf [Accessed 27 September 2014].

Franklin, A., 1999. *Animals and modern cultures: a sociology of human–animal relations in modernity*. London: Sage.

Gallant, K., Arai, S., and Smale, B., 2013. Serious leisure as an avenue for nurturing community. *Leisure Sciences*, 35 (4), 320–36.

Gillespie, D.L., Leffler, A., and Lerner, E., 2002. If it weren't for my hobby, I'd have a life: dog sports, serious leisure, and boundary negotiations. *Leisure Studies*, 21 (3–4), 285–304.

Haraway, D., 2008. *When species meet*. Minneapolis, US: University of Minnesota Press.

Ingold, T., 2000. *The perception of the environment: essays on livelihood, dwelling and skill*. London: Routledge.

Irvine, L., 2004. *If you tame me: understanding our connection with animals*. Philadelphia, US: Temple University Press.

McShane, C. and Tarr, J.A., 2007. *The horse in the city: living machines in the nineteenth century*. Baltimore, US: The Johns Hopkins University Press.

Nosworthy, C., 2013. *A geography of horse-riding: the spacing of affect, emotion and (dis)ability identity through horse–human encounters*. Newcastle-upon-Tyne, UK: Cambridge Scholars Publishing.

Raisborough, J., 1999. Research note: the concept of serious leisure and women's experiences of the Sea Cadet Corps. *Leisure Studies*, 18 (1), 67–71.

Stebbins, R.A., 1992. *Amateurs, professionals and serious leisure*. Montreal, Canada: McGill-Queen's University Press.

Stebbins, R.A., 2001. Serious leisure. *Society*, 38 (4), 53–7.

Stebbins, R.A., 2004. Introduction. *In:* R.A. Stebbins and M. Graham, eds. *Volunteering as leisure/leisure as volunteering: an international assessment*. Cambridge, UK: CABI Publishing, 1–12.

Stebbins, R.A., 2012. *The idea of leisure: first principles*. New Brunswick, US: Transaction Publishers.

Thompson, K., 2011. Theorising rider–horse relations: an ethnographic illustration of the centaur metaphor in the Spanish bullfight. *In:* N. Taylor and T. Signal, eds. *Theorizing animals: re-thinking humanimal relations*. Leiden, Netherlands: Brill, 221–53.

Wilkie, R.A., 2010. *Livestock/deadstock: working with farm animals from birth to slaughter*. Philadelphia, US: Temple University Press.

5 '... and horses'

The affectionate bond between horses and humans/gods in Homer's *Iliad*

Tua Korhonen

The first artwork of European literature, the *Iliad*, tells about a period of eight weeks in the war between the Trojans and the Achaeans (Greeks) at the end of the Trojan War. But the best-known episode of this decade-long war is in fact absent from the *Iliad* – namely the Trojan horse. The Greek troops gave this gigantic hollow wooden structure as a treacherous gift to the Trojans, which led to the sack of Troy. But why did the Greeks donate a gigantic *horse*?

According to another, only fragmentarily extant, epic about the destruction of Troy (*Iliou persis*), the Trojans assumed the wooden horse to be a kind of farewell gift from the departing Greeks, a votive offering to the goddess Athena (Burkert 1972: 178–81; Franko 2005–06). Besides the fact that enormous size was usually attached to divinity, and horses symbolized high status and wealth in general (only the rich could afford to keep horses), the city of Troy was famous for its horses (Delebecque 1951: 39). The epithet of the city was *eupôlos* 'abounding in horses' (Homer, the *Iliad* 5.551, 16.576) and the Trojans were frequently called *hippodamoi* ('horse taming', ibid. 4.352). Moreover, the first Trojan kings had had a herd of divine, immortal horses, a pair of whose semi-divine offspring was stolen by the Greek hero Diomedes during the war, an incident which forms an important part of the plot in the fifth book of the *Iliad*.

Containing nearly sixteen thousand lines, the *Iliad* also includes passages which illuminate the human–horse relationship. Do the heroes of the *Iliad* tender the same kinds of emotions towards their horses as we do? And what about the archaic audience of Homer? It is, of course, hazardous to make conclusions about archaic ways of thinking, or ways of treating horses, from the way horses are treated in a stylized fiction like the *Iliad*. The Homeric epics are, at least, known for their basic 'realism', which pertains in a way to their depictions of animals. Although Achilleus' divine horse Xanthus once even talks to his master (Homer, the *Iliad* 19.408–17), this incident is treated as an exception, and the mythic aspects are kept in the background. Greek mythology contains winged horses like Pegasus as well as anthropophagic horses, but Homeric horses are clearly recognizable as embodied horses, flesh and blood.

In the following, I will first explore the depicted human–horse relationship and horse care before pondering the possible affectionate relationship between horses and humans and, to a lesser extent, between horses and gods in the *Iliad*. By 'affectionate bond' I simply mean some kind of positive emotional attachment, reciprocal or not, which may be discerned in the human–horse encounters in the *Iliad*.

Chariot horses and Homeric warfare

Although the *Iliad* was put into a written form only in the eighth or seventh century BCE, it contains elements from the preceding early Iron Age civilization (the so-called Dark Age) and even from the Mycenaean, Bronze Age civilization of the second millennium BCE. Conscious archaizing elements, which the composer of the *Iliad* has utilized in order to create the atmosphere of the heroic past, are, for instance, the use of horse chariots in the war and the episode of human sacrifice at the end of the *Iliad* (Homer 23.171–6). The latter occurs in Patroclus' funeral, where Achilleus places not only enemies, nine Trojan prisoners, but also animals on his friend's funeral pyre. Besides sheep and cattle this animal sacrifice consists of four horses and two of Patroclus' companion dogs.[1] The belief system behind this custom was probably the same as in Egypt: these animals were intended to accompany the hero's soul to Hades. Sacrificing horses, donkeys and dogs did take place during the Bronze and early Iron Age Greek funerals according to the archaeological evidence, which is quite extensive about horses. Many horse burials contain one or more pairs of horses, usually together with a chariot or cart (Richardson 1993: 187–8).

The two- or four-horse-drawn chariots were used in early Iron Age warfare, and driving chariot horses – not riding on horseback – was a basic element of horsemanship in the *Iliad*. Although aristocratic warriors fought on foot, they were transported to the battlefield by horse chariots, which were sometimes used as a fighting vehicle as well (Latacz 1977: 215–23; Crouwel 2012: 55). Harnessing and unharnessing the chariot horses, driving, feeding and taking care of horses as well as stealing the enemy's horses or horse chariots are frequent actions of the narrative of the *Iliad* – they are so-called formulaic-type scenes (Delebecque 1951: 180–1; Janko 1992: 46). Horses which were known to be excellent were precious war booty and thus they might function as an inducement to heroic exploits, as when Hector persuades his soldiers to spy on the Greek camp by promising the enemy's best horses as a reward for their services (Homer, the *Iliad* 10.306, 330). The importance of horses is seen in the catalogue of Greek chieftains and their troops, where the best two pairs of horses, their outward appearance and excellent speed, are mentioned (ibid. 2.760–70). Sometimes the great warriors were recognized by their horses in the chaos of the battlefield (ibid. 5.183, see also 23.455).

Besides being part of the plot, horses also occur in the animal similes of the *Iliad*. The simile is a peculiar Homeric technique, an elaborate, digressive comparison, where humans and the human world are compared to animals and the natural world. Among the horse similes is one which describes a runaway stallion:

As when a horse that is kept at the manger and fed full with barley
breaks its tether and gallops exultantly, hoofs drumming,
over the plain, since its habit is to bathe in the waters
of a sweet-flowing river; it holds its head high, and its mane

flows about its shoulders, and confident in its splendour
its legs carry it easily to the haunts and pastures of horses;
so Paris, Priam's son, strode down from high Pergamus,
shining brightly in his armour like the beaming sun, and laughing
aloud as his swift feet carried him along (Homer, the *Iliad* 6.506–14).[2]

Besides being a simile, this is a eulogy to the dynamic vitality of a horse whose existence is not seen as human-related. The stallion is seen as its own entity, as having a life of its own. It is questionable whether the horse in this simile is a riding horse or a chariot horse.[3] Otherwise, the horses in the *Iliad* are, as already mentioned, almost exclusively chariot horses so that the word *hippos* meant both a horse and a horse chariot in the *Iliad* (Delebecque 1951: 141–2).

Managing horses in the middle of the battle is described as demanding great skills of the charioteers. Chariots were used for pursuit or flight or they remained stationary while the warrior battled (Greenhalgh 1973: 58–9). The close contact of horses and humans is expressed by a recurrent expression which refers to a situation when the warrior can feel the horse's breath on his shoulders or back (Homer, the *Iliad* 13.385 and 17.501, see also 23.380); the charioteer kept his chariot close to the warrior for a speedy escape. The essential quality of chariot horses was reaction speed: they had to speed 'in pursuit and retreat, galloping this way and that across the plain', as the Greek warrior Diomedes put it (ibid. 8.106–7).

The excellence of horses was measured in horse races (Homer, the *Iliad* 9.124–7). Although a war epic, the *Iliad* contains an intense description of a chariot race, namely, the first contest in the funeral games arranged in honour of Patroclus at the end of the *Iliad*. Even the gods, Apollo and Athena, interfere in the games, just as they have interfered in the war. Athena is on the side of Diomedes and fills his horses with vital force (*menos*) (ibid. 23.390, 23.400). Not only are the chieftains and gods described as being excited by the race but even horses are mentioned to be eager to run, like Agamemnon's mare Aethê, whom his brother, Menelaus, drives (ibid. 23.300, 23.525). Old Nestor's son Antilochus addresses his father's horses in the heat of the race urging them to go past Agamemnon's Aethê:

Press on, the pair of you; now for an all-out effort!
I am not urging you to compete with those there, the horses
of Tydeus' war-minded son [Diomedes], on whom Athena has
just now bestowed speed and has put glory into Diomedes.
Faster now, try to catch the horses of Atreus' son [Menelaus]! Do not fall
behind, and do not let Aethê who is only a mare, pour scorn
over the pair of you. Why are you lagging, my champions *[pheristoi]*?
I tell you this plainly, and it will surely be fulfilled:
you will get no more care from Nestor, shepherd of
the people, but he will kill you at once with the sharp bronze
if your slipshod ways mean that we win a lesser prize.
So off you go after them, run as fast as you can!
My part will be to fashion and devise some stratagem

to slip past them in a narrow place; I shall not miss the chance (Homer, the *Iliad* 23.403–16).

Antilochus is proud of his father's horses: they are the best or bravest ones, *pheristoi* – translated here as 'my champions', due to the vocative form – and the honour of the house is upon their placing in the race.[4] The threatening words ('Nestor… will kill you… if') reflect the harsh language of warriors, full of intimidation, common in heated exchanges in the battle scenes of the *Iliad* (ibid.).[5] The address is an expression of the common exertion of a human and horses but also a distribution of work: horses should run as fast as they can whereas Antilochus, the driver, should 'devise some stratagem' (ibid.).

The Homeric world also includes divine and semi-divine horses. The immortal gods own immortal horses, which eat ambrosia (Homer, the *Iliad* 5.777) and are able to cover long distances at a single stride and at the speed of thought (ibid. 5.769–72). Despite their divine nature, they function as 'working' horses, as chariot horses. The gods and goddesses not only drive their chariots majestically, like Zeus and Hera in the air (ibid. 8.374–83, 5.768–75) and Poseidon on the surface and in the depths of the sea (ibid. 13.23–30), but sometimes also unharness and harness them with precious equipment (ibid. 5.720–32, 8.41–4). Although immortal, the Olympian gods are not invulnerable: Aphrodite and Ares are both wounded in the battle (ibid. 5.352–69, 5.858–9). Vulnerability also concerns divine horses: angry Zeus at one time threatens to lame the divine horses of disobeying gods, break their chariots and wound the gods themselves (ibid. 8.402, 416).

Mortals may also own divine horses; they are not, however, described as having the capability of moving at the speed of thought like the gods' horses. The Trojan royal house had received its divine horse breed from Zeus with the exchange of prince Ganymede, whom Zeus desired (Homer, the *Iliad* 5.265–7). According to another story, the second king of Troy owned three thousand beautiful horses and Boreas, the North Wind, falling in love with them, metamorphosed himself into a black stallion. The result of this union was twelve divine horses, which were so swift that they were able to run on the surface of the sea (ibid. 20.220–30) (Harrison 1991: 253). This is told by Aeneas, whose father Anchises took his mortal horses near these royal, immortal ones, which resulted in a semi-divine breed. Aeneas took two of these semi-divine horses into the battlefield and it was this pair which the Greek hero, Diomedes, managed to snatch (ibid. 5.268–72, 295–6, 319–27, and 23.348) and with whom, later on, he won the horse race in Patroclus' funeral games (ibid. 23.500–13). On the Greek side, too, there were divine horses or horses of divine origin.[6] Achilleus received his immortal horses Xanthus and Balius from his father Peleus, who obtained them as a wedding gift from the gods when marrying the goddess Thetis (ibid. 23.276–8).

Taking care and taking into consideration

Notwithstanding Antilochus' lengthy address quoted above, addressing one's horses is not very common in the *Iliad*.[7] Hector, the son of Priam, the king of Troy, addresses his four steeds by name – 'Xanthus and you, Podargus, Aethon and bright Lampus' (Homer, the *Iliad* 8.185)[8] – and urges them on before fighting by reminding them of the good care they have had from his wife, Andromache, who served the horses even before her husband and gave them honey-sweet unground corn mixed with wine (ibid. 8.184–97). The symbols of human culture, grain and wine, are here the diet of horses, which equates horses with humans. Elsewhere more common horse food is mentioned, such as clover and wild parsley (ibid. 2.776). When the troops come back to their camps or citadels, the first thing is to unharness the horses (ibid. 8.50, 8.504). Another detail of the care, mentioned, however, only once, is anointing horses' mane with olive oil after washing them with water (ibid. 23.281–2).

Odysseus takes the sensitivity of horses into consideration in the episode when he steals Rhesus' excellent white horses. While sleeping his first night at Troy the Thracian prince Rhesus, who had come to help the Trojans, was killed with twelve of his warriors by Odysseus' companion Diomedes. Odysseus draws the corpses aside 'with this plan in his mind *[thymos]*, that the fine-maned horses might/pass easily through the camp and not tremble in their hearts *[thymos]*/as they trod on dead men' (Homer, the *Iliad* 10.491–3). The passage shows how, due to their recent arrival at the battlefield, prince Rhesus' horses have not yet become used to the smell of blood and dead bodies. It is noteworthy that the poet applied the same word, *thymos*, to denote the planning human mind of Odysseus and the mind ('heart' in the translation) of the horses. *Thymos* is one of the many Homeric words referring to the mind or the soul (or the life force) and it was used, as this passage shows, both for humans and non-human animals alike (Delebecque 1951: 157).

The Lydian prince Pandarus had also come to help the Trojans, but he left his precious horses at home. However, he refuses to drive Aeneas' chariot because he knows that the familiar driver can make horses not only run faster but also manage them if they go wild. Thus Pandarus says to Aeneas: 'I am afraid that if they cannot hear your voice they will grow/restive and take fright, and refuse to carry us out of the battle' (Homer, the *Iliad* 5.233–4). The voice of the familiar charioteer is thus seen as an important factor in the human–(chariot) horse relationship.

As partners in human war, the agency of horses may even cause humans to change their war tactics. This happens clearly once in the *Iliad*. The Trojans fail to continue conquering Greek ships because horses refuse to run over the ditch the Greeks have built around their seashore camp. Hector urges the troops to go, but 'not even his swift-footed horses would attempt it for him, but stood/whinnying loudly at its very edge; the wide ditch/terrified them, and it was not easy to jump or to/cross' (Homer, the *Iliad* 12.50–4). Horses are thus depicted voicing their fear by whinnying loudly.

Descriptions of horses in the *Iliad* thus include references to horses' reactions and bodily involvement in human war or games. Horses get tired, their bodies are covered with foam, dust and blood, and they run wild, and tramp over fallen soldiers.[9] They get injured and killed, too. Dying is an important recurrent theme of the *Iliad*. The battle scenes are full of short, detailed descriptions of deaths of otherwise insignificant human warriors, which makes death a singular event. Everyone dies individually.

There are even two short descriptions of lethal injuries suffered by horses: one is Nestor's, another Achilleus' horse, which is driven by his companion Patroclus (Homer, the *Iliad* 8.80–6, 16.467–9). Both are so-called trace horses, which are not yoked. Their function in Homeric warfare is still unclear, but they were probably used as backup horses (Janko 1992: 337, 379). The Trojan prince Paris hits Nestor's horse with an arrow to the head, at the specific place which the narrator knows is fatal for horses:

> it was hit on top of the head, where a horse's mane starts
> to grow upon its skull, and it is a most vulnerable point.
> The arrow sank into its brain, and it reared up at the pain,
> and reeling from the bronze it stampeded the other horses (Homer, the *Iliad* 8.83–6).

The pain thus makes the horse stagger and it throws other horses into confusion. Later on it is told that Nestor's horses (but most probably not this lethally injured one) are transported from the field to safety (ibid. 8.113–4).

Pedasus, Achilleus' mortal trace horse, which is called 'faultless' (Homer, the *Iliad* 16.152), belongs to Achilleus' famous team consisting of a pair of divine horses (Xanthus and Balius). A Trojan warrior aimed his spear at Patroclus, but missing him, instead struck the horse Pedasus in the right shoulder: '[...] and it screamed as it gasped its life *[thymos]* away/and fell bellowing in the dust, and the life *[thymos]* flew from it' (ibid. 16.467–9).

Pedasus' injury seems to be mild but by using the word *thymos* twice it is confirmed that the horse actually suffers death and, what is more, dies like a human: its soul leaves it.[10] Earlier in the narrative, it has been reported that Achilleus had taken Pedasus as war booty from the nearby sacked city (Homer, the *Iliad* 16.153–4). Thus, Pedasus' 'origin' is told as is often the case in the short descriptions of dying warriors (e.g., ibid. 8.303–5).

On the basis of all this, horses seem to have a surprisingly pervasive and profound presence in this first masterpiece of Western literature. Their fate in war is equated in many ways to humans – horses in the Trojan War are afraid, get injured and die, but also eat and get rest like human warriors. The war is a source of suffering and toil for humans and horses alike. This is sometimes stated by using the conjunctive *kai*, 'and'. War is said to be the 'ruinous toil of men – and horses' (Homer, the *Iliad* 17.400). Achilleus is mentioned in his massacring rage to have killed men – *and* horses (ibid. 21.521), although in the detailed account of the massacre only the killing of humans is described (ibid. 20.384–489).[11] However, there are no depictions of pity or sorrow for injured or dead horses anywhere in the *Iliad*. This does not necessarily mean that there was no affection but that the display of emotion is restricted due to the so-called heroic code or because of the literary genre. Edouard Delebecque, in still the most useful study of horses in the *Iliad*, noticed (1951: 9) that the horses in the *Iliad* have psychological nuances. But are there also psychological nuances in the depictions of the human–horse relationship?

Affectionate human–horse relationships in the *Iliad*

As mentioned earlier, the Lydian prince, Pandarus, had left his horses at home against his father's wishes. He wanted to spare (*pheidesthai*) them. Pandarus is slightly bragging about his wealth (for which the Lydians were famous), saying that in his father's halls he has no less than eleven fine, brand-new chariots and for each of them a pair of horses. However, it is clear that while speaking of sparing them he is not referring to his beautiful new chariots but his horses. He is afraid that his horses, which were used to eating barley and wheat, would not get as good food as at home (Homer, the *Iliad* 5.192–203). This may not just be the idea of protecting his property, his expensive creatures, or merely bragging how he serves his horses expensive food, but an expression of genuine affection as well. But still, it is not an explicit expression of emotional attachment.

We may say that the value system of Homeric heroes, the heroic code, included the good care of horses with the reservation that the gentle treatment of horses may be expected even for purely practical reasons. Horses – as well as common soldiers – were taken good care of in order to make them able to endure great exertion in fighting and in games (Homer, the *Iliad* 2.381–83; Lilja 1974: 74, 77). They were valuable as good workers. That the work of war demanded time-consuming training for chariot horses was a good enough reason to try to spare them. The more expensive horses were treated as precious commodities, valuable objects of pride and possession.

There are quite a few occurrences of pity (*eleos*, *oiktos*) for the deaths and injuries of fellow soldiers in the *Iliad*. This 'Homeric pity', as Glenn W. Most (2004: 62) has argued, is usually combined with feelings of anger and resentment, a desire to avenge the death of fellow soldiers. It thus promotes their will to continue the fight. However, there are even more deaths and sufferings of individuals, which are pictured to happen without any expression of pity. The 'modernistic' matter-of-fact descriptions of death and injuries are part of the poetics of the *Iliad*. The pace of the narrative does not allow the expression of emotions in the heat of the battle contrary to, for instance, lyric poetry or epigrams.[12] Therefore, although the warriors are not depicted as expressing pity for a suffering horse, it is not necessarily for the reason that they cannot identify with the suffering of the non-human other, their animal partners in war.

But did it belong to the heroic code to openly display one's emotions to one's horse? Characters in the Homeric epics seem as such to be emotional and have a tendency to act impulsively. However, Homeric characters and their affections are, as Michael Silk (1987: 83) has argued, described and defined mainly by their public status or outward features – not as having emotions peculiar to their personality. Moreover, the characters seem to be strangely self-centred. When Andromache learns of Hector's death and anticipates the sack of Troy, she seems to mourn less the fate of Hector than her own fate as a future slave (Homer, the *Iliad* 24.723–45).

Despite the absence of horsemen openly displaying their affection to their horses, there is one memorable scene in the *Iliad* where somebody pities the fate of two war horses. When Achilleus' divine horses, Xanthus and Balius, mourn the death of Achilleus' dear friend Patroclus, their sorrow does not arouse the sympathy of the surrounding fighting humans but the supreme god Zeus, who pities (*eleein*) them and regrets ever giving them to humans. After that Zeus even rescues the horses by breathing valour into them, which makes them able to move and, finally, after some incidents, arrive safely back at the Greek camp (Homer, the *Iliad* 17.441–58). It is noteworthy that the target of the supreme god's pity is the divine, not the mortal, horses, which underlines the distinction between immortality and mortality – not between human and non-human.

The catatonic sorrow of the horses is an anticipation and projection of Achilleus' forthcoming grief. Xanthus and Balius stood out in the fight, as still as a grave pillar, weeping and letting their manes be soiled – like humans pouring dust over their heads when mourning (Edwards 1991, 106).[13] Thus, Achilleus' horses show their loyalty to their master by weeping over the death of Patroclus, who has been their gentle driver and care taker, too. But it is their relationship with Achilleus which later becomes the focus and where the affective aspect of the human–horse relationship is most poetically expressed.

For one thing, Xanthus and Balius are hardly able to be driven by any other except Achilleus himself (Homer, the *Iliad* 10.402–4, 17.76–8). They thus obey only the orders of Achilleus, who has a semi-divine origin himself. When Achilleus finally decides to join the battle wanting to take revenge for Patroclus' death, he blames – that is, he addresses – his horses for not bringing his friend's dead body back from the battlefield. Xanthus answers that Patroclus' death was an act of the gods as will be the near death of Achilleus himself; this time, however, they bring Achilleus safely back to the Greek camp (ibid. 19.400–24). As a divine horse, Xanthus knows the will of the gods and therefore sees his master's future. But Achilleus already knows that he will eventually be killed in Troy and will not be able to go back home. Therefore his answer expresses annoyance: 'Xanthus, why do you prophesy my death? There is no need' (ibid. 19.420).

Animals functioned as omens and signs of future events and, moreover, there were prophetic speaking animals in folktales (Edwards 1991: 283). However, I suggest that the scene of the speaking Xanthus can be interpreted as a sublime fictionalization of the sensitive bond between Achilleus and his precious and unique horses, the wedding gift of the gods to his father. Achilleus is not surprised when Xanthus answers him, as though a speaking horse were an ordinary event. It is the narrator who explains that the ability of speech is a gift of the goddess Hera (Homer, the *Iliad* 19.407). We may interpret Xanthus' words as projections of Achilleus' thoughts while speaking to his horses and imagining what these divine animals would answer. He knows that he is going to die.[14] Or, that Achilleus is able to understand his horses in this uncanny way is an expression of his liminal state. Or, it is a poetic fictionalization of an everyday phenomenon of human–animal communication. In a continuous human–animal relationship, communication is always two-way and sensitive owners or trainers can 'read' their animals through the animals' body language and tone. Others, like Nestor's son Antilochus and Hector, address their horses at length, as quoted above. But they do not listen to what the horses might say as Achilleus does. Is it possible that in the Homeric world only a divine horse was thought worth listening to?

However, the vulnerability and mortality of the Homeric horses participating in the Trojan War is equated with that of humans' participation. They share the common harsh fate of war. They are like ordinary soldiers who obey orders and die without anybody pitying their fate. Yet, there is at least one person who seems to feel pity and sympathy for all mortals – both men *and* horses alike in the *Iliad* – namely the poet. All in all, the poet of the *Iliad* depicts horses as living creatures with psychological nuances, but he also points to their role as symbolically important and prestigious animals. However, whether prestigious or not the Homeric horses have hardly any other choice than to follow their masters to war as fated by the gods.

Notes

1 Sheep, cattle, dogs and humans are mentioned as being killed before being thrown on the pyre. Horses are only 'flung' onto the pyre, which cannot, of course, be done with living horses. Achilleus is said to do this 'with loud groans'.
2 All translations, except individual terms, are by Anthony Verity (2011).
3 The simile occurs for the second time at 15.263–8 (Homer, the *Iliad*), where Hector returns to the battle.
4 Both Antilochus and Menelaus urge or rebuke their horses, which are said to run faster thence (Homer, the *Iliad* 23.403–16 and 23.443–5).
5 There is no reference that Antilochus was going to carry out his threat, although he won only the second prize. Note the scorn for the female gender, which transgresses the boundaries of species ('Aethê who is only a mare') (Homer, the *Iliad* 23.408).
6 In passing, Nestor also mentions the famous divine horse Arion (or Arêion) (Homer, the *Iliad* 23.346–7). On Arion, see Heath (1992: 397).
7 For other addresses to horses, see Homer, the *Iliad* 8.184–97, 19.400–3, 23.443–5, and 16.684 and 17.431.
8 For horse names in the *Iliad*, see Delebecque (1951: 146), and in Greek culture in general, Calder (2011: 45–6).
9 See, for instance, Homer, the *Iliad* 5.233, 5.296, 6.39–42, 8.136, 11.127–29, 11.282–3, 16.379–82, 20.495–7, 23.507–8. Horses could also be injured during the games, in horse races (ibid. 23.341, 23.435), and accidentally injure humans.
10 As, for instance, Urs Dierauer (1977: 9) has pointed out, dying in the *Iliad* is depicted not simply by the dichotomy: human *psyche* – animal *thymos*. See also Rahn (1950–54: 434–5).
11 Hera mentions the toil and trouble which the war has caused both her and her horses (Homer, the *Iliad* 4.26–7).
12 Epigrams were the most popular poetic form during Hellenistic times (c. 330–30 BCE), containing funerary epigrams to animals, such as war horses (Herrlinger 1930).
13 Horses do, in fact, have tear glands, which keep their eyes clean. Achilleus' horses mourn Patroclus' death also in Homer, the *Iliad* 17.425–8 and 23.279–85.
14 Also Thetis, Achilleus' divine mother, had warned her son of near death (Homer, the *Iliad* 18.95–6).

References

Burkert, W., 1972. *Homo Necans: Interpretationen altgriechischer Opferriten und Mythen.* Berlin: Walter de Gruyter.

Calder, L., 2011. *Cruelty and sentimentality: Greek attitudes to animals, 600–300 BC*. Oxford, UK: Archaeopress.

Crouwel, J.H., 2012. *Chariots and other wheeled vehicles in Italy before the Roman Empire*. Oxford, UK: Oxbow Books.

Delebecque, E., 1951. *Le cheval dans l'Iliade suivi d'un lexique du cheval chez Homère et d'un essai sur le cheval pré-homérique*. Paris: Librairie C. Klincksieck.

Dierauer, U., 1977. *Tier und Mensch im Denken der Antike*. Amsterdam: Grüner.

Edwards, M.W., 1991. *The Iliad: a commentary. Vol. V: books 17–20*. Cambridge, UK: Cambridge University Press.

Franko, G.F., 2005–06. The Trojan horse at the close of the '*Iliad*'. *The Classical Journal*, 101 (2), 121–3.

Greenhalgh, P.A.L., 1973. *Early Greek warfare: horsemen and chariots in the Homeric and archaic ages*. Cambridge, UK: Cambridge University Press.

Harrison, E.L., 1991. Homeric wonder-horses. *Hermes*, 119 (2), 252–4.

Heath, J., 1992. The legacy of Peleus: death and divine gifts in the *Iliad*. *Hermes*, 120 (4), 387–400.

Herrlinger, G., 1930. *Totenklage um Tiere in der antiken Dichtung*. Stuttgart, Germany: Kohlhammer.

Homer, 2011. *The Iliad.* Trans. Anthony Verity. With an introduction and notes by Barbara Graziosi. Oxford, UK: Oxford University Press.

Janko, R., 1992. *The Iliad: a commentary. Vol. IV: books 13–16*. Cambridge, UK: Cambridge University Press.

Latacz, J., 1977. *Kampfparänese, Kampfdarstellung und Kampfwirklichkeit in der Ilias, bei Kallinos und Tyrtaios*. München, Germany: Beck.

Lilja, S., 1974. Theriophily in Homer. *Arctos*, 8, 71–8.

Most, G.W., 2004. Anger and pity in Homer's *Iliad*. *In:* S. Braund and G.W. Most, eds. *Ancient anger: perspectives from Homer to Galen*. Cambridge, UK: Cambridge University Press, 50–75.

Rahn, H., 1950–54. Tier und Mensch in der homerischen Auffassung der Wirklichkeit. *Paudeuma*, 5, 277–97 and 435–80.

Richardson, N., 1993. *The Iliad: a commentary. Vol. VI: books 21–24*. Cambridge, UK: Cambridge University Press.

Silk, M., 1987. *Homer: the Iliad*. Cambridge, UK: Cambridge University Press.

PART II

Mapping human–animal spaces

Relationality

6 Re-reading sentimentalism in Anna Sewell's *Black Beauty*

Affect, performativity, and hybrid spaces

Jopi Nyman

Anna Sewell's horse novel *Black Beauty* (2012; first published in 1877) is a classic work addressing human cruelty and the insignificance of equine life in Victorian Britain. Through its portrayal of the life story of Black Beauty from idyllic childhood to old age, narrated, as its original subtitle puts it, as an 'autobiography' told by the horse himself, the novel criticizes contemporary animal-keeping practices and ideologies where horses often had a merely instrumental value. What is particularly significant in Sewell's novel is that the episodic and chronicle-like narrative presents a didactic message of humane treatment of animals by using sentimentalist discourse to the extent that the novel was defined by its American publisher as 'The Uncle Tom's Cabin of the horse' (Angell 2012: 176). The relationship between the two texts has been addressed in criticism and has led to several readings contributing to an understanding of *Black Beauty* as a narrative of slavery (see Gavin 2012: xxi; Ferguson 1994). While Robert Dingley (1997: 242) finds the similarities 'more than coincidental', in Peter Stoneley's (1999: 72) view the two texts are linked in their way of addressing the black body and the desire that it represents, challenging the maintenance of white female identity. Both *Uncle Tom's Cabin* and *Black Beauty*, through their graphic descriptions of broken families, violated bodies, and monolith ideologies, rely on the conventions of the slave story in order to appeal to the emotions of the readers. In so doing, each text presents an emotionally appealing narrative of wrongs committed by those in power against those unable to defend themselves successfully.

In the case of *Black Beauty*, this emphasis on affect is, however, more than a mere strategy seeking to support the ideological thesis of the novel or an expression of Sewell's underlying Quaker values (see Hollindale 2000: 97–8). Rather, the critical re-reading of *Black Beauty* presented in this chapter aims at reading its sentimentalism as a form of renegotiating the human–animal relationship amidst Victorian modernity. In contrast to the pleas for human-centred humanity promoted in *Uncle Tom's Cabin*, *Black Beauty*'s sentimental discourse telling of the abused horse – which made it the key text for the emerging animal protection movement – foregrounds the animal as a site of affect. In so doing it replaces the question of race with that of species and calls for an animal-centred reading. What is at stake here is that the animals represented in *Black Beauty* enter a space conventionally defined as human and dominated by human values, and in so doing they redefine its preferred roles and identities. This means that the novel's animal representation is involved in the redefinition of dichotomies such as the human and the animal as discussed in traditional Western philosophy since Descartes (see Nyman 2003: 13–14). As a sign of this, the novel's act of providing the 'dumb beast' with a voice and history is not merely an act of anthropomorphizing the animal. Rather, the representation of Black Beauty as a speaking animal capable of story-telling aims at challenging the alleged human monopoly over reason and human identity as the natural way of being in the world.

As I will show in my reading of the novel, the animal presences itself in space that should be shared rather than dominated by humans. The chapter will contextualize the novel within the discourse of sentimentalism promoted by the sentimental novel in Britain and the United States since the late eighteenth century as it places the novel in the context of affect theory. My analysis of the novel will start with its representation of the suffering animal body and the emotional work it incites and will continue with the performance of animal identity in diverse spaces. As I will show, *Black Beauty*'s sentimental power derives from its interrogation of conventional animal identity and its locations.

Approaching the sentimental

By placing the suffering animal at the centre of the narrative, *Black Beauty* engages in the tradition of sentimentalism, often considered formerly to be a second-rate form of writing by popular or women writers aiming at, to use the words of Janet Todd (1986: 2), 'the arousal of pathos through conventional situations, stock familial characters and rhetorical devices' and 'an emotional, even physical response'. Critics locate the origins of sentimentalism in eighteenth-century Britain and its 'cult of sensibility', aiming to provide models for behaving and responding emotionally in particular social situations through fiction (Todd 1986: 4). The emotional aspect is also evident in the shifting meanings of the term sentiment: while in the mid-eighteenth century its meanings stressed rationality and judgment, its nineteenth-century derivatives add a further meaning of feeling and are often pejorative (Banfield 1997: 2–3). As Banfield's (1997: 4; emphasis original) discussion of *The Comprehensive English Dictionary* (1864) shows, at that point '*Sentimentalism* and *sentimentalist* appear as derivatives which imply moral diminishment'. The contemporary usage is aptly defined by June Howard (1999: 76) as follows: 'when we call an artifact or gesture sentimental, we are pointing to its use of some established convention to evoke emotion; we mark a moment when the discursive processes that construct emotion become visible'.

As a mode of expression, sentimental literature is characterized by strong emotional elements evident in its content and its attempt to influence the reader. Todd (1986: 4) characterizes the work of a sentimental text as follows: it 'moralizes more than it analyses and emphasis is not on the subtleties of a particular emotional state but on the communication of common feeling from sufferer or watcher to reader or audience'. In a similar vein, Glenn Hendler (2001: 3), drawing on the thinking of Adam Smith, finds the emotion of compassion, or sympathy, at the core of the sentimental novel, as the reader's identification implies that he or she 'imagine[s] oneself [...] in another's position'. This capacity of sentimentalist texts to link readers with texts, to create communities, is what Nicola Bown (1997: 4) sees as a key motivation for examining such often neglected works today: 'Sentimental art and literature invites us sympathetically to share the emotional world of those distant from us in time and circumstance'.

Since the focus of the sentimental discourse in *Black Beauty* is on the suffering of the animal that is to be shared by the reader, the novel can be examined within the framework of compassion, a term that has a long and chequered history. Recently philosophers and affect theorists have sought to re-evaluate the concept. According to Lauren Berlant (2004: 4; emphasis original), compassion is more than a theoretical concept as it gains its meaning in social practice '*as an emotion in operation*'. In her view there is an imbalance of power between the sufferer and the person noticing it: 'the sufferer is *over there*. You, the compassionate one, have a resource that would alleviate someone else's suffering' (ibid.; emphasis original). For Berlant (ibid.), the feeling of compassion is problematic, as it may be used for various political purposes to develop appropriate feelings, 'to feel right'. Philosopher Martha Nussbaum has a more positive view of compassion, a feeling that, in her view (1996: 28), serves as 'a bridge' connecting people with their communities. For Nussbaum (1996: 29), compassion is a moral way of responding that involves a recognition of both pity and suffering, not irrationally but through the feeling of compassion, which 'involve[es] thought or belief' (Nussbaum 1996: 31). In Nussbaum's (1996: 35) view the suffering of others matters to us if there is a sense of shared community.

Compassion, then, expresses solidarity and is a means of social inclusion by learning of the suffering of others by 'cross[ing] boundaries of class, nationality, race, and gender' (Nussbaum 1996: 50), as is made possible in such practices as multicultural education (ibid.). According to Kathleen Woodward (2004: 72), Nussbaum's position represents a 'liberal' rather than a 'conservative' view of compassion, a view that understands 'pain and suffering' as its basic elements. Such a position recognizes the 'unequal' distribution of power and the different positions of the sufferer and the witness, and also problematizes the question of whether moral responses may lead to social justice (Woodward 2004: 73). Woodward (2004: 71), however, compares Nussbaum's view with that of Berlant, in whose view narratives of sentimentalism are problematic because of an unresolved contradiction between the private and the political. To quote Berlant (1998: 641): 'when sentimentality meets politics, it uses personal stories to tell of structural effects, but in so doing it risks thwarting its very attempt to perform rhetorically a scene of pain that must be soothed politically'. In her discussion of Berlant's position, Woodward (2004: 72) notes that its critique is targeted at the project of optimism evident in sentimental literature. While sentimental literature cultivates a position of feeling rather than leading to action, the 'postsentimental' literature of James Baldwin and Toni Morrison, in Berlant's view, combines sentimentalism with such emotions as outrage and critique that have a clearer cognitive component (Woodward 2004: 72–3). To be effective, compassion needs to recognize its non-universality, with each case addressed contextually (Woodward 2004: 70).

While *Black Beauty* is a highly sentimental text attacking utilitarian and instrumentalist views of animals, a closer reading of selected episodes in the novel will reveal the ways in which it negotiates between human and animal spaces, and breaches conventional boundaries. As I will argue, its foregrounding of the animal body in various positions – such as moving and performing, subject and object, agent and victim – is what makes it a powerful text of compassion where the onlooker learns about the animal and its predicament. Following Samuels's view (1992: 3) suggesting that the body – 'nation's bodies and the national body' – is a central feature in sentimental writing, I will approach the novel through a discussion of the various ways in which *Black Beauty* relies on the trope of the *animal* body, both as an object of the human gaze and as a site of non-human agency and performativity. What such doubleness contributes to is a negotiation of animal identity that challenges the conventional binary divisions and categories of modernity by showing the intertwinement of human and animal identities and the possibility of sharing spaces.

Moving body I: moving the reader(s)

While sentimental abolitionist writing foregrounded the suffering black body to support its critique of slavery (see Samuels 1992: 5), Sewell's representation of the animal body utilizes similar scenes of physical violence in order to show the cruelty of contemporary animal treatment. In nineteenth-century Britain, the horse carried various and, occasionally, contradictory meanings, ranging from the promotion of gendered and patriarchal power to female docility and submission (Dorré 2006: 10–11). In *Black Beauty*, both aspects are present: while some of its representations such as elegant carriage horses serve to construct elite class identities, on other occasions the body of the horse is likened to other marginalized identities such as those of gender and race. These latter representations contribute to the construction of compassion in the narrative, and a particularly striking example concerns the practice of 'breaking' – a practice whose representation in Dorré's (2006: 110–11) view resembles the act of rape, in which the body is violated and penetrated. The descriptions of the introduction of the bit into the horse's mouth emphasize the discomfort of the experience. Sewell (2012: 14-15) describes the moment as one when

> a great piece of cold hard steel as thick as a man's finger [is] pushed into one's mouth, between one's teeth and over one's tongue, with the ends coming out of the corner of your mouth, and held fast there by straps over your head, under your throat, round your nose, and under your chin […].

This passage can be compared with Ginger's narrative of the same experience that constructs her experience as one in which she is forced to submit to the will of her human masters and lose her 'liberty' – a term that links the novel with the discourse of slave narratives (see Sewell 2012: 25).

To support its message of the omnipresence of animal abuse, Sewell's novel features several further scenes where horses are 'flogged' and 'whipped' mercilessly by their owners and drivers (e.g., 2012: 26, 46, 66). As is typical of sentimentalist writing, as proposed by Todd, the detailed representations are often combined with moral messages that both promote humane treatment and also urge those witnessing such episodes to take a stand and act, and not only to feel. What the text aims to view sympathetically is the animal itself, whose suffering is made visible by emphasizing the pain that it experiences. Rather than viewing the animal as an Other, the text, through its strategy of providing the horse with a voice, seeks to construct a community to enable the feeling of compassion. When the reader encounters the pain of the horse, the text promotes an emotional response; when the text critiques the cruelty of such practices and presents alternative modes of behaviour; its compassion has a cognitive and political dimension.

The significance of compassion as a moral emotion is central to the novel's pedagogical project. The malicious ploughboy Dick is punished for throwing stones at the horses. Young Joe Green sees a violent carter who abuses his two horses by 'sw[earing] and lash[ing] most brutally' (Sewell 2012: 66) and reports it to the owner of the brickfield, and, as the following passage shows, upon coming across similar event, the Bertwick Park coachman John Manly reacts in a similar manner. While the pony manages to escape from the situation, Manly mentions it to the boy's father, who promises to punish his son. Sewell (2012: 46) describes the treatment of the horse in detail:

> at some distance we saw a boy trying to leap a pony over a gate; the pony would not take the leap, and the boy cut him with the whip, but he only turned off on one side; he whipped him again, but the pony turned off on the other side. Then the boy got off and gave him a hard thrashing, and knocked him about the head; then he got up again and tried to make him leap the gate, kicking him all the time shamefully, but still the pony refused.

In this novel, scenes of animal abuse serve two purposes: they incite affective responses and aim at promoting more humane rational and moral action because of their educational function – which appears to have been Sewell's original aim (see Gavin 2004: 186). In so doing the novel promotes a sense of liberal compassion as articulated by Nussbaum and Woodward, arguing that to be effective, emotional responses should be combined with cognitive reasoning. As the text shows, education and knowledge are needed to improve the condition of horses in Victorian Britain. This is particularly evident in the way it suggests that the younger generation plays a major role in this issue, as seen in the child and adolescent characters of the novel and also in its reception.

To underline the point, the novel not only suggests that humans are responsible for the well-being of animals but that this is to be taught to children through daily actions and choices, as is shown on several occasions. Little Joe Green learns to take care of horses through trial and error, and the London cab driver Jerry Barker's children Dolly and Harry help their father in taking care of the horses. The following passage telling of Harry's willingness and hard work while his father is on his sickbed is particularly illustrative:

It was late the next morning before any one came, and then it was only Harry. He cleaned us and fed us, and swept out the stalls; then he put the straw back again as if it was Sunday. He was very still, and neither whistled nor sang. At noon he came again and gave us our food and water; this time Dolly came with him; she was crying, and I could gather from what they said, that Jerry was dangerously ill, and the doctor said it was a bad case (Sewell 2012: 146).

In addition to showing Harry's ability to take care of the horses, it emphasizes his sense of responsibility during the family's ordeal in gendered terms. While Dolly cries openly, Harry carries on with his duties regardless of his own suffering. Through such representations, the novel seeks to promote preferred forms of behaviour as well as to instil a particular sense of morality into the minds of its readership. In this sense *Black Beauty* is clearly linked with the conventions of nineteenth-century children's literature where potentially dangerous children are transformed through '(Evangelical) reformative education and guidance' [into] 'well-mannered and content subject[s]' (Vallone 1991: 74).

The novel's power to promote the well-being of animals through education was also recognized in the early reviews of the novel. While the *Eastern Daily Press* suggested that 'all boys and girls' with contact with horses 'should possess this volume' (Gavin 2012: xv), *The Essex Standard, West Suffolk Gazette, and Eastern Counties Advertiser* (1878: 6) recognized its 'intent of inculcating kindness to horses not only as a matter of principle, but also as the best way of insuring a good disposition in an animal on whose tractability we so great depend'. In the view of the anonymous reviewer, its values are not only literary but also educational: 'it is an excellent book to put into the hands of stable boys, or any who have to do with horses' (ibid.). The role of the work as a means of teaching a non-instrumental way of relating to horses was quickly recognized, as a school edition was published as early as 1878 (Gavin 2012: xv). A review in the *Nottinghamshire Guardian*, in 1899, of a later sixpenny edition mentions its international reputation and praises the novel for its 'admirable lessons of kindness and humanity', as well as wishing that through this cheap edition 'its influence for good will be still more widely felt' (1899: 5).

This process of moral education also works at the discursive level of the Sewell novel, as suggested in Horst Dölvers's (1993) analysis. Dölvers (1993: 206) argues that the key words and concepts in the novel such as 'gentle' and 'good' have a double function: when denoting their objects as 'very good colts', they ascribe to them a moral value. In Dölvers's (1993: 207) view, this pedagogical strategy aims to promote 'Victorian Christian righteousness', which can be achieved by 'be[ing] loving and good, do[ing] your work willingly, [and] avoid[ing] bad habits'. Such a reading can be supplemented by linking it with Nussbaum's (1996: 50) view arguing that compassion can be learnt through reading, a view that can also be traced in the reviews of the novel discussed above. The novel also underlines, by using Joe Green and Harry as examples, that such a development is possible for everybody, even the most brutal of boys. It has, indeed, been noticed that on several occasions in the novel, and especially in

Merrylegs's story, that boys relate to horses as if they were 'steam engine[s] or [...] thrashing machine[s], and can go on as long as and as fast as they please; they never think that a pony can get tired, or have any feelings' (Sewell 2012: 33). While Merrylegs's way of teaching them is to 'r[i]se up on [his] hind legs and let [them] slip off' (Sewell 2012: 33), the teaching of appropriate emotions is a social task set for the text. That boys in particular would learn proper ways of treating horses by reading the book was recognized in its contemporary marketing. Gavin (2012: xv–xvi) quotes the following unattributed view from 1882: 'Miss Sewell teaches so pleasantly, that even the most abrupt of boys must be won over to follow the good and to shun the evil course in their conduct towards the dumb servants of man'. Consequently, a good human–horse relationship demands that the horse is recognized as a sentient being that is able to respond to the human. Good horsemanship is based on mutual understanding and trust rather than on violence and abuse, as seen in Sewell's (2012: 134) description of the 'young coster-boy' and his pony selling vegetables in Jerry Barker's street:

> he had an old pony, not very handsome, but the cheerfullest and pluckiest little thing that I ever saw, and to see how fond those two were of each other, was a treat. The pony followed his master like a dog, and when he got into his cart, would trot off without a whip or a word, and rattle down the street as merrily as if he had come out of the Queen's stables. Jerry liked the boy, and called him 'Prince Charlie', for he said he would make a king of drivers some day.

While promoting the need to recognize the status of the horse as a being needing adequate living conditions, rest, and care, the novel also reveals some of the reasons for the maltreatment horses in a world containing little social justice. Rather than juxtaposing the value of human and non-human life, the novel contextualizes the problem of promoting equine well-being in terms of class by presenting a situation where the well-being of the horse and that of its driver are interdependent. This is the case of Seedy Sam, 'a shabby, miserable-looking driver', whose horse appears 'dreadfully beat' (Sewell 2012: 128). As drivers like Sam do not own their horses but rent them, they have to work exceptionally hard to support their families, which in turn leads to animal abuse:

> you must put your wife and children before the horse, the masters must look to that, we can't. I don't ill-use my horse for the sake of it; none of you can say I do. There's wrong lays somewhere, – never a day's rest – never a quiet hour with the wife and children (Sewell 2012: 129).

Even though Seedy Sam's view appears to work against the promotion of horse welfare, it contextualizes the issue of abuse in terms of class and suggests that a more humane treatment of drivers and a fairer economy would also work for the benefit of the horses.

However, the novel's representation of the animal body as a non-human Other serving as a passive object of compassion in need of human intervention and help is under negotiation in the novel because of its critique of the conventional human–animal division. As the following section shows, *Black Beauty* reveals how the categories of human and animal, self and other, and reason and non-reason are challenged. In a similar vein, by representing space as shared by humans and animals, the novel challenges the alleged human mastery pertinent to discourses of modernity.

Moving body II: performing animal spaces

Rather than mere silent objects of human maltreatment, the horses represented in *Black Beauty* are also agents able to perform their identity in human space. Rather than a fault or an implausibility, the anthropomorphism of the narrative, seen in its speaking horses, their values and views, can be seen as contributing to the feeling of compassion. This is also what Schuurman and Leinonen (2012: 62) suggest in their statement that '[a]nthropomorphism can [...] also be investigated as a method of increasing mutual understanding between humans and animals'. From this perspective, *Black Beauty* appears as an intercultural text seeking to make the culture of the horses available to humans and to show how their identity, while different from that of the humans, shares sites and spaces with the latter. This project is evident in the peritexts of the novel: the title page of the first edition of the novel inscribes Anna Sewell not as the author of the novel but as its translator: 'Translated from the Original Equine, by Anna Sewell' (2012: 3). This act of translation, of making the world of the horses available to humans, tells of the novel's attempt to construct a community allowing for compassion. Unlike humans, the horses of the novel understand the languages of the humans and while often disagreeing with them, they defer to their ascribed status, aware of the fact that resistance may lead to punishment or worse. As Merrylegs puts it:

> if I took to kicking, where should I be? Why, sold off in a jiffy, and no character, and I might find myself slaved about under a butcher's boy, or worked to death at some seaside place where no one cared for me (Sewell 2012: 33).

The view of inequality between humans and horses is well established and crystallized in the words of Sir Oliver, a wise old horse at the Gordon's: 'We horses must take things as they come, and always be contented and willing so long as we are kindly used' (Sewell 2012: 35).

While Sir Oliver's view suggests that horse identity is something ascribed and normative, fixing discursively on the horse as the one 'used', the novel also shows that identity is performed in various spaces, offering the horse an opportunity to counter the script. Such spaces can be understood as animal spaces, a term defined by the geographers Philo and Wilbert (2000: 14) in the following manner:

> in such cases [...] it is animals themselves who inject what might be termed their own agency into the scene, therefore transgressing, perhaps even resisting, the human placements of them. It might be said that in so doing the

animals begin to forge their own 'other spaces', countering the proper places stipulated for them by humans, thus creating their own 'beastly places' reflective of their own 'beastly' ways, ends, doings, joys and sufferings.

The following discussion will address four distinct spaces of animal performance in the novel: the paddock, the horse fair, the battlefield, and the modern city. As these spaces are parts of the episodic narrative structure of the sentimental novel, the vignette-like structure of the novel serves to provide a variety of locations for the performance of animal identity, not only to emphasize Black Beauty's and Ginger's downfall, as Hollindale (2000: 103–6) has suggested. My reading of the performance of identity is based on the theoretical work introduced by Judith Butler and developed by other scholars. Butler (1990: 24) argues that identities such as gender are not expressions of any pre-existing self but they come into existence when they are performed in the public sphere: 'the substantive effect of gender is performatively produced and compelled by the regulatory practices of gender coherence'. Identity, in Butler's (1990: 140; emphasis original) view, is 'a *stylized repetition of acts*': when a new performance cites previous discourse differently, its effect may be subversion and change. In other words, identity is doing rather than being. Butler's views have been widely debated and applied in contemporary cultural studies. For instance, in their application of Butler to human geography, Nicky Gregson and Gillian Rose (2000: 441) suggest that spaces are also performative: the identities and relationships of participants create new spaces that may be in a critical relationship with those presented in previous discourses and scripts. With particular reference to human–animal studies, Birke *et al.* (2004: 168) argue that analysis of performativity may open up new perspectives by challenging 'the human/animal divide'. Performativity may serve to explain the ways in which human and animal both participate in the construction of their relationships and also generate new phenomena ('choreographies') together (Birke *et al.* 2004: 170–1). Such performances, Birke, Bryld, and Lykke (2004: 176; emphasis original) suggest, involve several actors and their interaction: 'there are three kinds of performativity here – of animality, of humanness [sic], *and of the relationship between the two*'.

Performativity plays a significant role in *Black Beauty*. The novel presents several forms of identity performance that are not exclusive but foreground various aspects of the identity of the horse. These performances can be examined as spaces or sites. The first space to be addressed is that of Home Paddock, where all of the horses at Birtwick Park are able to gather together and become involved in mutual communication. This orchard, like other similar sites in the novel, is an animal space in the sense defined by Philo and Wilbert (2000): it provides animals with opportunities for physical activities unrestricted by humans. But more importantly, it is a space of reflection for story-telling and the sharing of memories. This space of narrative interaction constructs a sense of cohesion for the horse community. The stories told to each other by Merrylegs, Ginger, and Sir Oliver (see Sewell 2012, Chs. VII–X) reveal their life histories and define the human–horse relationship from a non-human perspective. As a site, the paddock provides a safe space that is somewhat similar to Black Beauty's childhood home, 'a large pleasant meadow' (Sewell 2012: 9). To quote Sewell (2012: 24):

> The grass was so cool and soft to our feet; the air so sweet, and the freedom to do as we liked was so pleasant; to gallop, to lie down, and roll over on our back, or to nibble the sweet grass. Then it was a very good time for talking, as we stood together under the shade of the large chestnut tree.

By defining the paddock as a space of reflection, I aim to emphasize its characteristic discursive activity, consisting of the horses' discussions and definitions of their ideals and their past relationships with humans. Particularly negative encounters include the story of the cutting of Sir Oliver's tail (see Sewell 2012: 35–6), Ginger's history of abuse, including her violent breaking in by the drunken Samson, a violent groom, and the painful bearing rein – which is compared to the corsets of the time by Dorré (2006: 110–1) – necessitated by London fashion but making her mouth bleed (see Sewell 2012: 29–30). Upon hearing the story, young Black Beauty appears to take over the role of a non-believer and may thus be compared to a reader pondering on the truth of the story: 'of course I knew very little then, and I thought most likely she made the worst of it' (Sewell 2012: 31). In this way Sewell's (2012: 35) novel contrasts young Black Beauty's idealism with the harsh realities of horse abuse and problematizes his notion of the good horse as a reliable servant of his master whose 'greatest pleasure' is to be 'saddled for a riding party': 'It was so cheerful to be trotting and cantering all together, that it always put us in high spirits'. This change is seen in the fact that upon hearing tales of cruelty from old Sir Oliver, Black Beauty feels 'a bitter feeling toward men rise up' (Sewell 2012: 37). This space of story-telling, however, appears free from violation against animals, which the horses also comment on by pointing out that their owner and his men 'are good masters and good grooms' (ibid.) who refrain from practices causing pain. This comment indicates that these horses do not dream of a natural space, a space without humans, but as domesticated horses their identity is intertwined with that of the humans. Liberty, as Black Beauty puts it, rather than separation from humans and the tasks they set ('I know it must be so' [Sewell 2012: 23]), means regular access to exercise and the pleasures of the paddock. In sum, the performance of horseness in the space of reflection produces communally critical non-human reflections on the human–animal relationship that seek to counter the established discourses and practices of animal treatment. While the ideal of the good horse is present in the discussion, the life stories reveal why not all of the horses may comply with it and demand that trust be reconstructed with 'patience and gentleness, firmness and petting' (Sewell 2012: 31).

The second space of the horse fair functions as a contrast to the sense of community and agency characterizing the paddock. In this space the performance of the horse is regulated by humans whose motives are merely financial and aim at profiting at the expense of animals. Rather than having intrinsic value, here horses are commodities to be sold to the highest bidder, which links the novel with the representation of the slave markets in *Uncle Tom's Cabin*. Here, unwanted, aged, and sick horses, like Black Beauty with his 'broken knees', constitute the greatest number of items to be sold. They are described in a manner demanding a compassionate response and issuing a call for improved horse welfare:

> But round in the back ground, there were a number of poor things, sadly broken down with hard work; with their knees knuckling over, and their hind legs swinging out at every step; and there were some very dejected-looking old horses, with the under lip hanging down, and the ears laying back heavily, as if there was no more pleasure in life, and no more hope; there were some so thin, you might see all their ribs, and some with old sores on backs and hips. These were sad sights for a horse to look upon, who knows not but he may come to the same state (Sewell 2012: 103).

The references to the horse body as one of illness and deterioration can be examined in the context of Victorian discourses on health and illness as outlined by Athena Vrettos. According to Vrettos (1995: 2), the period's emphasis on the physiological body as a site of illness speaks of an attempt to 'understand and control [the] world through a process of physiological and pathological identification'. The 'instability' generated by illness appeared problematic because of 'its capacity to reconfigure conceptions of the self' (Vrettos 1995: 3). While in this case the ailing body is that of an Other, of an animal, the sentimentalist narrative suggests that the condition of being unable to care for oneself is something that may concern anyone 'dejected' in unfortunate circumstances and lead to a lessening of their value. This commodified animal identity is also characterized by a further fear of loss generated by this threatening space that the non-human actors remain unable to control, as the interaction of the space is regulated by human salesmen prone to lying and 'trickery' (Sewell 2012: 103). Escape and upward social mobility are firmly restricted to those 'in their prime, and fit for anything' (ibid.). What emerges in this performance of horse identity is a sense of its insecurity and its dependence on human beings to provide the conditions for a life worth living.

The third space for the performance of horse identity is the battlefield, where horses perform gallantry and heroism in the name of the nation, as is evident in the story of Captain, a former military horse but now a London cab horse. By linking the horse with the national project, the novel aims to promote several ideas: the horse as a useful and necessary actor in the making of Empire; its reliability as a partner and the depth of the bond; and its gendered performance of gallantry and heroism. Unlike the abused Ginger, Captain does not complain about his past experiences but relates them to Black Beauty in a way that shows his commitment to the national cause:

> we always liked to hear the trumpet sound, and to be called out, and were impatient to start off, though sometimes we had to stand for hours, waiting for the word of command; and when the word was given we used to spring

forward as gaily and eagerly as if there were no cannon balls, bayonets, or bullets (Sewell 2012: 109–10).

The main emphasis in Captain's story is on his version of the infamous battle of Balaclava in October 1854, known best for the suicidal charge of the Light Brigade, commemorated in Tennyson's classic poem (Gavin 2004: 120). The novel shows how the horse – then known as Bayard (see Sewell 2012: 196) – participates in the attack against the enemy and loses his 'dear master', an officer who has told him just before the charge that 'we'll do our duty as we have done' (Sewell 2012: 110).

This use of the first person plural signifies, of course, the rhetoric of a unified nation when confronting an enemy, but it also constructs a smaller dyad of the rider and the horse who work together. This is an example of the human–animal hybrid discussed by Birke, Bryld, and Lykke (2004: 169–70), who see it an example of 'animaling' – a process in which the emergent joint identity is different from that of the human or the animal. This idea of human–animal hybridity plays a significant role in the performance of horse identity in Sewell's novel. For instance, Captain's narrative suggests that the relationship demands the contribution of both partners: when the officer dies, the riderless Captain, 'without a master or a friend, [...] alone on that great slaughter ground' (Sewell 2012: 111), trembles and nearly loses his nerve, until a new rider emerges to continue the battle. This loss of control appears to have less to do with a view defining horses as 'inferior' and 'irrational' animals and humans as their 'natural' masters than with the recognition of the particular features of the human–horse hybrid. It is evident that *Black Beauty* is aiming at replacing conventional views separating humans and animals from each other with a relational view that emphasizes their intertwinement and shared achievements.

Such a view is clearly promoted in the animal performance central to the novel's fourth space, the modern city, where Black Beauty works together with the cab driver Jerry Barker. This performance, more than the others in the novel, is based on the hybrid human–horse dyad and promotes a view of the animal as having access to knowledge and agency. In the city, it appears, a successful cab driver needs a good awareness of the conventions of the urban. From the perspective of the horse, however, such knowledge is not possessed by humans only, but also by horses. The following passage shows how it is through their mutual understanding, trust, and shared actions that success is possible:

It is always difficult to drive fast in the city in the middle of the day, when streets are full of traffic, but we did what could be done; and when a good driver and a good horse, who understand each other, are of one mind, it is wonderful what they can do. I had a very good mouth – that is, I could be guided by the slightest touch of the rein, and that is a great thing in London, amongst carriages, omnibusses, carts, vans, trucks, cabs, and great waggons creeping along at a walking pace; some going one way, some another, some going slow, others wanting to pass them, omnibusses stopping short every few minutes to take up a passenger, obliging the horse that is coming behind, to pull up too, or to pass, and get before them; [...] If you want to get through London fast in the middle of the day, it wants a deal of practice (Sewell 2012: 114–5).

As the passage shows, this space of the modern city is a hybrid space containing both humans and non-humans, whose values and projects, as the dyad of the cab driver and his horse shows, can benefit from each other. This performed space is not a pure one but, as a result of the presence of various actors, it is, to use the terms of Gregson and Rose (2000: 442), 'threatened, contaminated, stained, enriched by other spaces'. In devising a new and jointly performed identity, the novel successfully subverts conventional discourses of positing humans and non-humans as separate and even opposite categories – here they share the same space and, following the idea put forward by Gregson and Rose (2000: 441), bring that space into being in their joint performance. This emphasis on hybridization appears to be one of *Black Beauty*'s major achievements: while seemingly a pedagogical and sentimental text, its performance of animal identity rests on a reconstructed relationship between the human and the non-human, who appear to need each other to be able to prosper. As a result, modernity's opposition between humans and animals, reason and feeling, dissolves and is replaced with an idea of their mutual intertwinement.

Conclusion

This chapter has sought to re-read Anna Sewell's sentimental text *Black Beauty* within the contexts of affect and performativity. I have suggested that the novel's use of the suffering animal body plays a significant role in the construction of compassion, inciting sympathy for animals. This sympathy, however, is not merely affective but also cognitive as it contributes to historical discourses of moral education. The findings of the chapter are also relevant from the perspective of human–animal studies as it constructs shared spaces where animal spaces flow into and contaminate formerly uniformly human-dominated spaces. Within such spaces, humans and horses may perform together and form hybrid and intertwined identities (cf. Birke *et al.* 2004). In so doing, *Black Beauty* replaces conventional understandings of the separateness of humans and animals with an understanding of their togetherness.

What distinguishes *Black Beauty* from many other sentimental texts is its particular construction of affect as political. Rather than suggesting that concern for animals is a private response, constantly it promotes a more general and humane way of animal treatment through appeals to education and in so doing participates in the construction of what Glenn Hendler (2001: 12) has labelled as 'public sentiment'. What this means is that by voicing sentimental responses to animal suffering in the public sphere and by locating the animal issue as a problem of national concern, the novel calls for political change through feeling. In so doing, Sewell's novel opens up new strategies of identification and problematizes conventional ideas of humans, animals, and their place(s).

References

Angell, G., 2012. What is overloading a horse, and how proved? *In*: Anna Sewell. *Black Beauty*. Ed. A. Gavin. Oxford, UK: Oxford University Press, 176–77.

Banfield, M., 1997. From sentiment to sentimentality: a nineteenth-century lexicographical search. *19: Interdisciplinary Studies in the Long Nineteenth Century* [online], 4. Available from: www.19.bbk.ac.uk/index.php/19/article/view/459/319 [Accessed 19 February 2015].

Berlant, L., 2004. Introduction. *In*: Lauren Berlant, ed. *Compassion: the culture and politics of an emotion*. New York: Routledge, 1–13.

Berlant, L., 1998. Poor Eliza. *American Literature*, 70 (3), 635–68.

Birke, L., Bryld, M., and Lykke, N., 2004. Animal performances: an exploration of intersections between feminist science studies and studies of human/animal relationships. *Feminist Theory*, 5 (2), 167–83.

Bown, N., 1997. Introduction: crying over Little Nell. *19: Interdisciplinary Studies in the Long Nineteenth Century* [online], 4. Available from: www.19.bbk.ac.uk/index.php/19/article/view/453/313 [Accessed 19 February 2015].

Butler, J., 1990. *Gender trouble*. London: Routledge.

Dingley, R., 1997. A horse of a different colour: *Black Beauty* and the pressures of indebtedness. *Victorian Literature and Culture*, 25, 241–51.

Dölvers, H., 1993. 'Let beasts bear gentle minds': variety and conflict of discourses in Anna Sewell's *Black Beauty*. *AAA: Arbeiten aus Anglistik und Amerikanistik*, 18, 195–215.

Dorré, G., 2006. *Victorian fiction and the culture of the horse*. Aldershot, UK: Ashgate.

Ferguson, M., 1994. Breaking in Englishness: Black Beauty and the politics of gender, race, and class. *Woman: A Cultural Review*, 5 (1), 34–52.

Gavin, A.E., 2004. *Dark horse: a life of Anna Sewell*. Phoenix Mill, UK: Sutton Publishing.

Gavin, A.E., 2012. Introduction. *In*: Anna Sewell. *Black Beauty*. Ed. A. Gavin. Oxford, UK: Oxford University Press, ix–xxvii.

Gregson, N. and Rose, G., 2000. Taking Butler elsewhere: performativities, spatialities and subjectivities. *Environment and Planning D: Society and Space*, 18, 433–52.

Hendler, G., 2001. *Public sentiments: structures of feeling in nineteenth-century American literature*. Chapel Hill, US: The University of North Carolina Press.

Hollindale, P., 2000. Plain speaking: *Black Beauty* as a Quaker text. *Children's Literature*, 28, 95–111.

Howard, J., 1999. What is sentimentality? *American Literary History*, 11, 63–81.

Nottinghamshire Guardian, 1899. Literature and Art. *Nottinghamshire Guardian*, 10 June, p. 5. Issue 2820.

Nussbaum, M., 1996. Compassion: the basic social emotion. *Social Philosophy and Policy*, 13 (1), 27–58.

Nyman, J., 2003. *Postcolonial animal tale from Kipling to Coetzee*. New Delhi: Atlantic.

Philo, C. and Wilbert, C., 2000. Animal spaces, beastly places: an introduction. *In*: C. Philo and C. Wilbert, eds. *Animal spaces, beastly places: new geographies of human–animal relations*. London: Routledge, 1–34.

Samuels, S., 1992. Introduction. *In*: S. Samuels, ed. *The culture of sentiment: race, gender, and sentimentality in 19th-century America*. Oxford, UK: Oxford University Press, 1–8.

Schuurman, N. and Leinonen, R.-M., 2012. The death of the horse: transforming conceptions and practices in Finland. *Humanimalia: A Journal of Human/Animal Interface Studies* [online], 4 (1), 59–82. Available from: www.depauw.edu/humanimalia/issue%2007/schuurman-leinonen.html [Accessed 19 February 2015].

Sewell, A., 2012. *Black Beauty*. Ed. A. Gavin. Oxford, UK: Oxford University Press.

Stoneley, P., 1999. Sentimental emasculations: *Uncle Tom's Cabin* and *Black Beauty*. *Nineteenth-Century Literature*, 54 (1), 53–72.

The Essex Standard, West Suffolk Gazette, and Eastern Counties Advertiser, 1878. Literature. *The Essex Standard, West Suffolk Gazette, and Eastern Counties Advertiser*, 24 August, p. 6.

Todd, J., 1986. *Sensibility: an introduction*. London: Methuen.

Vallone, L., 1991. 'A humble spirit under correction': tracts, hymns, and the ideology of evangelical fiction for children, 1780–1820. *The Lion and the Unicorn*, 15, 72–95.

Vrettos, A., 1995. *Somatic fictions: imagining illness in Victorian culture*. Stanford, US: Stanford University Press.

Woodward, K., 2004. Calculating compassion. *In*: L. Berlant, ed. *Compassion: the culture and politics of an emotion*. New York: Routledge, 59–86.

7 Seeing the animal otherwise

An Uexküllian reading of Kerstin Ekman's *The Dog*

Maria Olaussen

'When does something begin? Where does a tale begin?'[1] (Ekman 2010: 1–2). These questions form the opening sections of Kerstin Ekman's novel *The Dog* and draw our attention to the fact that we rely on ideas of time and place in the telling of a story. The answers to the first question are hesitant: 'It doesn't begin. There's always something else before it. It begins the way a stream starts as a rivulet and a rivulet starts as a trickle of water in the marsh. It's the rain that makes the marsh water rise' (Ekman 2010: 1). In a similar manner the answer to the question 'when does something end?' (Ekman 2010: 129) towards the end of the novel, relies on the tropes of recurrent natural cycles:

> Perhaps never. There's always something else after it. Hunting days follow with sun and strong wind, and lazy days of rest, head against forepaws, listening to the heavy rain streaming through the branches of the aspen and drumming on the tin roof (Ekman 2010: 129).

What follows in the closing section does, however, for the first time in the narrative, move our attention from the experiences of the dog towards the story about the dog, 'the man often told the story […] and he also spoke often of how' (Ekman 2010: 130), and the novel ends by further stressing the difference between the ending of a tale and the end of an event.

The Dog was originally published as *Hunden* in Swedish in 1986 and tells the story of a small puppy that gets lost in a wintry forest while following its mother. The dog learns how to survive in the forest but is found by a hunting party and slowly comes to trust one of the hunters and returns to human habitation to live as a hunting dog with him. The main focus of the novel is, however, on the life of the dog in the forest where the struggle for food and warmth in competition with and in fear of other animals come to occupy centre stage.

By foregrounding the narrative structure and the parameters of time and place needed for the construction of a tale, the novel, in my view, shows how these techniques, necessary as they are for human communication, remain inadequate when it comes to understanding and representing the recurrent and cyclical aspects of events in nature. In this sense the events that mark the beginning and the end of the story, the puppy getting lost and then returning to a life with humans, can be read as only one large cycle of departures and returns captured within the narrative. The ending stresses the tension between the end of the story and the fact that the dog continues to wait and 'remained alone in his waiting' (Ekman 2010: 133).

This tension between the anthropocentric and the animal experience is brought out through the use of a narrator who has the function of making animal experience intelligible within the conceptual world of the human. Narrative is, in David Herman's (2009: 2) definition, 'a basic human strategy for coming to terms with time, process, and change – a strategy that contrasts with, but is in no way inferior to, 'scientific' modes of explanation that characterize phenomena as instances of general covering laws'. In *Animal Stories: Narrating Across Species Lines*, Susan McHugh (2011: 5) similarly argues against 'traditional notions that aesthetic forms follow from scientific thinking about animals' and sees narrative ethology as a way of recognizing the centrality of different forms of species to storytelling processes. It is in this function as a narrative and a work of art that *The Dog* contributes to what Matthew Calarco (2008: 127) identifies as an alternative way of representing animals to that of scientific discourse:

> Animals are not viewed as fundamentally alien to human language and concepts but rather as coextensive with the scientific discourses that purport to describe them. Consequently, animal ethicists rarely make recourse to poetic, literary or artistic descriptions of animals – descriptions that might help us to see animals otherwise, which is to say, otherwise than the perspectives offered by the biological sciences, common sense, or the anthropocentric 'wisdom' of the ages.

This effort to move beyond the scientific, of grasping life at its own terms, can be traced to Martin Heidegger's (1995: 188) discussion in *The Fundamental Concepts of Metaphysics* of the representation of life that 'must be accomplished on the basis of the fundamental character of living beings themselves as something that cannot be explained or grasped at all in physico-chemical terms'. An approach which attempts to explain living substance through physics and chemistry is, according to Heidegger (ibid.), bound to fail by leaving us with life as an 'inexplicable residue'.

A similar preoccupation with life as understood through art constitutes one of the central challenges of Ekman's novel. As this chapter will show, based on careful observation of canine behaviour *The Dog* is, however, also an attempt to bring together art and science and make use of and produce knowledge in the widest possible sense.

Mud and sky

Kerstin Ekman is an award-winning Swedish novelist,[2] member of the Swedish Academy,[3] and author of 24 novels and works of non-fiction, of which several have been translated into English and several other languages.[4] She is known for her depictions of the forest regions of northern Sweden but also for the naturalist observations that permeate her novels. In a 2013 interview in the Swedish journal *10TAL,* in a special issue on *Klimatsorg* [Climate Sorrow],[5] she mentions *Masters of the Forest* (2008; originally *Herrarna i Skogen* [2007]) and *Se Blomman* [See the Flower] (Ekman and Eriksson 2011) as particularly important from an ecological perspective and describes *The Dog* as 'the impossible attempt to think beyond the human' (Jörneberg 2013: 40).[6] The novel *Forest of Hours* (1999; originally *Rövarna i Skuleskogen* [1988]) deals with the experiences of a troll as well as human–animal metamorphoses. Ekman traces her view of animals from early encounters with the poetry of Rainer Maria Rilke, especially his eighth Duino Elegy (1997 [1923]) and also stresses the connection between the development of European languages and the fact that what we now call Europe was once an area entirely covered with forest: 'We are of the forest, our languages and our knowledge have developed in the glades' (Jörneberg 2013: 38). This interconnection, Ekman insists, is fundamental to human existence and something often misrepresented as if 'nature' could be added to human life: 'we want it "with nature" as an added ingredient, not something that we ourselves are made of and live within' (ibid.). Although Ekman insists that being human means having a soul, a conscience and memory, the idea of the human as part of nature is present in the specific experience, in the walk through the woods through what 'smells, crackles, crunches and sways beneath your boots, and through what frightens you' (ibid.). This idea of human connectedness to a specific environment expressed through the walk in the woods is also present in Ekman's novel *Blackwater* (1996; originally *Händelser vid vatten* [1993]) where the loss of a forest area through clear-cutting is described through the paths that once ran through the forest. Memory is created through walking: 'Remembering right out into the stony scree. Not getting lost. Remembering with your feet. Not with a sick tumor called longing that reproduces images wildly and crudely and crookedly. No, foot memories, leg memories' (Ekman 1993: 403). The networks of paths are formed by both soles and hooves interacting with grass and scrub, retreating and bending, and form embodied memories in opposition to what Christopher Oscarson (2010: 14), in his reading of *Blackwater*, identifies as forms of 'environmental fundamentalism [that] ignore the entanglement of the subject within natural systems and within human economies and discursive practices'.

This idea of entanglement in Ekman's work is, however, modified by a focus on ideas of soul, consciousness and memory that become invested with ideals of traditional humanism.[7] In opposition to Donna Haraway's (2008: 3) 'I am a creature of the mud, not the sky', Ekman stresses the singularity of the human as reaching beyond nature while at the same time living, observing and being part of the natural world. Ekman's engagement with the traditions of Swedish and German Romanticism as well as the early naturalists in the essay collections *Masters of the Forest* (2008) and *Se Blomman* (Ekman and Eriksson 2011), together with her insistence on observing how 'the ground answers the foot', moves beyond the dichotomy between mud and sky. These texts combine technical, journalistic, literary, and personal styles to create a synthesis of literary texts, lived experience, and observation. In a discussion about *Masters of the Forest* with Anna Paterson

(2008b), Ekman points to the limitations of linear thinking[8] but both collections stress the importance of naming as a central form of species knowledge.[9]

In a discussion of consciousness Ekman finds it problematic to speak of a dog's sense of self-identity: 'Our dog cannot conceive of herself as "I". I don't think she even knows that she is a dog even though she knows the word. But she has strong feelings and memories but they need stimuli to come alive' (Jörneberg 2013: 38). Ekman's brief comment on *The Dog* as an impossible effort to think beyond the human could be seen as stemming from a tradition where thinking the animal subject starts with a preoccupation with the human. As Joanne Faulkner (2011: 77) argues, such 'self-knowledge – the ability to recognize that one is "human" – and self-mastery' are central to a 'founding act of separation' which inaugurates the human subject. Ekman's interest in trolls, *näcken*, and elves as situated beyond the human, lacking but longing for that which constitutes the human, is similarly an expression of a humanist concern. She traces her interest in the non-human to the stories about trolls told by her father, but also to the Swedish poet Erik Johan Stagnelius's poem 'Näcken' (1972) where the non-human being's longing for a soul is a central theme.[10] 'These stories and Stagnelius's poem have given me the feeling that there might be life without a soul and a longing for soul. That life without our religion, our conscience and all our intellectual constructions would be desolate' (Jörneberg 2013: 39).[11] In Krysztof Ziarek's (2008: 190) view 'this framing of the human within the relation between animality and humanity constitutes the originary gesture of humanism, which sets the scope, the limits and the categories within which the human can be at all comprehensible'. This 'human way' equals the traditional humanism of classical science which Agamben (2004: 40) describes as 'a single world that comprised within it all living species hierarchically ordered from the most elementary forms up to the higher organisms'.

Ekman's insistence on bringing that which she sees as existing beyond the human into literary form while keeping to a humanist framework foregrounds the idea of concurrences rather than disruption. *The Dog*, seen through the theories of *Umwelt*[12] by Jakob von Uexküll, opens up to an understanding of concurrent worlds where the legacy of humanism is retained even within an attempt to move beyond anthropocentric ways of perceiving non-human worlds.[13]

Concurrent worlds

In *A Foray into the Worlds of Animals and Humans: With a Theory of Meaning* Jakob von Uexküll (2010: 43) addresses himself to a reader who is to accompany him on an imaginary 'stroll [across] a flowering meadow'. Here, he explains, 'we make a bubble around each of the animals living in the meadow. The bubble represents each animal's environment and contains all the features accessible to the subject' (Uexküll 2010: 43). Although living in the same space and time, the subjects differ in what stimuli become accessible to them and this creates a new world within each bubble. Giorgio Agamben's (2004: 40) discussion of Uexküll stresses the radically altered view of the relation between a particular species and its environment that came with this idea of concurrent worlds:

Too often, he affirms, we imagine that the relations a certain animal subject has to the things in its environment take place and in the same time as those which bind us to the objects in our human world. This illusion rests on the belief in a single world in which all living beings are situated. Uexküll shows that such a unitary world does not exist, just as a space and a time that are equal for all living things do not exist.

Giorgio Agamben (2004: 40) further points to the relative insignificance of 'the *Umgebung*, the objective space in which we see a living being moving', in contrast to 'the *Umwelt*, the environment-world that is constituted by a more or less broad series of elements that he calls "carriers of significance" (*Bedeutungsträger*) or of "marks" (*Merkmalträger*), which are the only things that interest the animal'. This means that one and the same objective space, or *Umgebung*, contains several concurrent *Umwelten* or environment worlds, depending on the interaction between the animal and the carriers of significance in this world.[14] It is this interaction between the carriers of significance in the environment and the animal's receptive organs that Uexküll sees as a functional or musical unity:

> Everything happens as if the external carrier of significance and its receiver in the animal's body constituted two elements in a single musical score, almost like two notes of the 'keyboard on which nature performs the supratemporal and extraspatial symphony of signification,' though it is impossible to say how two such heterogenous elements could ever have been so intimately connected (Agamben 2004: 41).

Ekman's prose evokes a similar distinction between a space recognizable to her human readership and the environment created through the sense perception of the dog. The first hiding place of the dog, deep under the root of the spruce, is first described as a topological location, explaining how it remains cold and damp even at midday during a sunny day, and then followed by a description of cold, loneliness and hunger:

> He slept, and even when the sun was at its highest in the sky it never penetrated the root of the spruce where he was lying. This was on a northern slope, on a wooded rise behind the summer pasture. Faint with hunger, he lay with his muzzle tucked under his own backside to draw warmth from his body, from his own wet, matted coat and the cavity in which his heart fluttered like the wing of a bird in the cold and damp. It beat eagerly and wildly, throbbing, hungry for life, for warmth and kind voices, for milk, sunshine, tongues, fur, paws, belly and legs (Ekman 2010: 9–10).

It is significant that the beating of his heart is rendered through the metaphor of a bird's wing, existing independently of his volition or action and at the same time allowing for an extension of the meaning attached to this beating heart as a hunger for all that is suddenly missing in his life.

As Inga Pollman (2013: 2) argues, Uexküll's focus is on perception as vision and on a 'different paradigm of vision – a different *dispositif* – than that premised in the belief in a unified, objective world which can be known by an observer'. Uexküll's texts often address the reader as an observer, as in the opening of *A Foray into the Worlds of Animals and Humans* (2010) where the reader is invited along for the stroll but also in the closing section of *Umwelt und Innenwelt der Tiere*, entitled 'The Observer' [*Der Beobachter*]. Here Uexküll (1909: xxx) speaks about 'the pictures displayed' ['*Die vorgeführten Bilder*'] and begins his conclusion with the expression: 'If we look back at the *Umwelten* of the different animals that we have studied' ['*Werfen wir erst einen Blick zurück auf die Umwelten der verschiedenen Tiere, die wir betrachtet haben*'], thereby relying on the narrative strategy of showing rather than describing.

The challenge for Ekman lies in conveying the world of the dog in narrative form. In accounting for the sensory perception of the dog, the omniscient narrator occupies a position within the *Umwelt*-bubble[15] of the human while also, to a certain extent, accessing that of the dog. The tension between these two positions is created when the narrator addresses the reader in the opening section of the novel, then shifts to the perspective of a dog in accounting for life in the forest, and shifts back to the reader towards the end of the novel. Although the concept of *Umwelt* suggests a world distinction and could open up to forms of ontological metalepses with shifts between different worlds, there are no disorienting elements in the narrative that would suggest such radical movements.[16]

The shifts into the *Umwelt*-bubble of the dog are metaphorical and thematic rather than stylistic and also take place through movements from the topological to projective descriptions, where the experience of the dog is what determines the meaning of a particular place.[17] But rather than conveying this experience as a shift in perspective, the narrator focuses on the physical experience of a dog moving through the forest as in the following description:

> The moon was setting as he made his way back to the wooded area above the marsh. It was dark among the spruces and the snow wouldn't support his weight. Time after time he sank through until he finally curled up by a root, licking the salty tang of his blood from his paws. The woods were just coming to life as he fell asleep. At dawn one bird after another warbled tentatively in the dense, moist air. But he was sleeping (Ekman 2010: 27).

Here the narrator juxtaposes two worlds, the world of the dog and that of a human spectator watching the dog struggling through the snow. There is a strong emphasis on the difficulties experienced by the dog as the snow starts melting and he loses his way in a world that suddenly appears different. Whereas the reader is invited to understand this world through references to place as in the 'wooded area above the marsh' and to the impending dawn when the woods are 'just coming to life', the dog is described in his interaction with this environment, an interaction that here remains determined by the difficult movements through the forest and his exhaustion when finally reaching his destination.

The narrator thus presents the interaction between the dog and his *Umwelt* from within a perspective which remains intelligible to the human spectator. The focus on the dog is created through descriptions of what, in Uexküllian terminology, the dog's *Bedeutungsträger* are and the way in which the dog answers or reacts to these. The most important *Bedeutungsträger* in the narrative are the different scents. This is how the dog first learns to survive but also to recognize certain places and, finally, to structure his world around these places:

> The scent was coming from the snow. It made him start digging with his paws and poking with his nose. The crust was sharp but he broke through it quickly and the smell, the wonderful smell of food, wafted up. [...] He went on digging all the way down. Finally he sank his teeth into a frozen flank with a bristly hide. He had got through. It was food (Ekman 2010: 15).

This deepening awareness of the scents forms a pattern which shapes the narrative to suggest a sense of connection between time and place; the dog returns to the same places as before and recognizes this pattern of return through an encounter with his own scent. Recognition is thus framed as part of an ongoing and deepening interaction between the dog and his environment but the *Merkmalträger* and the *Bedeutungsträger* are here connected to the animal itself in a way that makes the symphony between dog and environment emerge out of his own movements through the landscape. To quote the novel:

> He came to a spot he recognised. It wasn't just one of the countless places where he caught a whiff of the restless phantom that was everywhere, his own scent. The silhouette of the grey building was familiar. He walked up to it, discovering the smell of his own urine on the wood of the door (Ekman 2010: 32).

The phantom of his own scent is thus an integral part of the specific world of the dog and memory is in part constituted through this recognition of a previous encounter manifested through the smell of his own urine.

Other ways of connecting scent to memory are created through what Ekman (2010: 41) terms 'indistinct trails of scent' which determine how and where he can choose a sleeping place: 'Sniffing around the place he'd slept, he wasn't sure what scents he picked up. Then he retreated, found another spruce or another pile of stones to crawl into. But he often returned to the old places that felt familiar, where he was on guard but not agitated'. At times the scents are detrimental to the dog as in situations of strong and overpowering smells:

> The grass was so full of strong smells that he sometimes had to raise his head to clear his nose. It smelled of yarrow about to bloom, a compact, heavily spicy scent. The mouldering humus was steaming, crawling with blind hard-shelled insects that ground their teeth, crisscrossed by fat, industrious bugs the thrushes could find by listening. He himself caught them only by chance when his sharp claws scratched the ground (Ekman 2010: 51).

The brief reference to the ability of the thrushes to find insects by listening suggests the existence of different environment worlds in the Uexküllian sense where the birds exist in interaction with the *Merkmalträger* of their *Umwelt*-bubble, differently constituted from that of the dog. A further suggestion of such a distinction between worlds is also found in the passage where the dog uses his sense of smell to find the birds' nests but also relies on their shrieks to guide him to the eggs. To quote Ekman:

> He always caused a commotion. Birds flew up in front of him with piercing shrieks that went on for a long, long time. That could mean eggs. He searched, nose to the grass, letting the shrieks guide him. When they grew loud and anguished he was close, when they died out he'd lost the trail (Ekman 2010: 57).

Here the reader watches and follows the dog in his attempts at learning how to find food and to recognize the meaning of the shrieks. The passages describing hunger and the search for food lead up to these incidents where the dog hunts successfully and thereby stays alive and where the agony of the birds are reduced to a disturbing element: 'Their chaos was between him and the forest' (Ekman 2010: 69). The idea of *Umwelt* as concurrent worlds is present in the narrative but in the same way as the narrator stays outside and watches the world of the dog, the dog also stays on the outside of the lives of other animals of the forest:

> All those living in dens and lairs around him had their own ways. From the moment daylight began filtering into their sleep until darkness fell and they tucked their beaks under their wings or curled their tails around their paws, each day was the same. They scurried up the same treetrunks and crept into the same holes. During the daylight hours they were constantly busy. Their world was familiar and they were on guard, for they all knew what was behind the tufts of grass and above the treetops, and what might be there (Ekman 2010: 40).

Both the narrator of this description and the dog are positioned on the outside of the world of these others and the fact that they are not referred to as animals' gestures towards a refusal of this type of categorization. 'Their ways' are observed by the dog but only they have the significant information of what constitutes their *Umwelt*. This doubleness, an effort to narrate the life of the dog from within the *Umwelt*-bubble of the human, is also evident when time is presented as a cycle of ever-returning events, extending from daylight to darkness, within a narrative that also uses expressions such as hours and days: 'That day he didn't hunt. He didn't recognise the forest around him. He was searching, but not for food; it was familiar places he was after, and the smell of his own markings' (Ekman 2010: 71).

Time, memory, waiting

An important aspect of Uexküll's theory of concurrent worlds concerns not only the idea of space as differently perceived by each subject but also the experience of time. Time exists only within the *Funktionskreis* of the subject and is interrupted, 'the world stands still' (Uexküll 2010: 52), whenever the animal finds itself in an environment devoid of the necessary stimuli. Uexküll (ibid.) concludes that 'Without a living subject, there can be neither space nor time'.

The opening of *The Dog* disavows openings by pointing to the fact that beginnings are arbitrary. This gesture towards the convention of determining the beginning of an event through a notion of time is also found in the description of the whereabouts of the dog. Encircled by the brown arm of a root under the wide skirts of spruce tree, the small dog is introduced as the subject of the narrative but the conventional expectation that the story contain a beginning, an explanation of how he 'had ended up under that spruce' is not possible, 'he didn't remember and he couldn't have told the tale' (Ekman 2010: 2). Instead we have a focus on the sensation of the dog, at this stage determined only through his need for food and warmth:

> There was nothing but a great big hole inside, a gnawing, a hunger for warmth and for the mild pungent sweetness that filled his mouth, and for his mother when she had come in from outside with strange scents in her fur, nipping at the scruff of his neck and licking the corner of his mouth (Ekman 2010: 2).

What is memory in the life of a dog? Ekman's novel explores the connections between stimuli in the dog's environment, links them to the dog's effect signs and develops this to include notions of memory and time. Hunger is represented as pain but also as memory where the hunger is for warmth and the mother's milk. The sound of a magpie similarly functions to connect a visual sensation to memories of the mother dog: 'Magpie chatter was followed by the mother dog's growling when the bright bird grew bold and came too close to their food. Insistent, penetrating sounds were also what he longed for: his mother and the food bowl' (Ekman 2010: 9). Traces of the mother keep recurring throughout the narrative, always in connection with external events in the forest which makes him anticipate the mother like in a distant howling: 'The howling was hoarse and piercing, carried by the wind from off the lake. It wasn't the right howl: his mother's' (Ekman 2010: 18). But with the passing of time, these traces are swept away in the same way that the traces of the mother dog that he followed on that fateful day when he got lost. It is significant that the loss of these actual traces in the snowy forest is represented through the imagery of a broom, a manmade device, in contrast to the nature imagery used to describe the impossibility of beginnings:

> A storm from the west is like a broom, a grey blast sweeping across lake and forests. Afterwards there is no trace of ski or snowmobile tracks, of animal or bird, no wads of snuff around the fishing holes, no bait, no blood. Everything is fresh, white and smooth (Ekman 2010: 7).

What appears to have one meaning suddenly changes to take on an entirely different one. This is a pattern which follows him throughout from that very first fateful shift in meaning: It is the misunderstanding of the mother dog that sees a rifle instead of an ice drill and sets off after the man on a fishing trip with the small pup following her, which leads the dog into the forest away from humans, house, mother, and siblings. With the lack of traces, the possibility of memory disappears and the dog will have to start anew with creating connections between what he experiences in his environment in the form of images, scents and his own sensations of fear, hunger, comfort, and cold. Although lost to the familiar, 'he roams through wisps of memory' (Ekman 2010: 85). To quote the novel,

> Day followed day and between them cold fragments of nights penetrated his sleep with the hooting of an owl or the snapping of a frozen branch. But he didn't connect the days in a series. His life and his memory were images upon images, fading in and out, scraps of days with bright skies, sharp scents to follow, disconnected cries wafting one by one through the woods until they attached to an image deep inside him (Ekman 2010: 18).

The early experience of winter months in the forest is primarily about survival. The dog needs food, warmth, shelter, and protection from predators but when this struggle is over and he learns how to adapt to his new environment, there is still a sense of something missing.

> Sleep and calm came from a full belly and the warmth of the sun, absorbed by his healthy, dry fur. But even then he was waiting.
> He didn't have an image or a name for what he was expecting. But he would recognise it when he heard it or caught its familiar scent. He lived in wait like a lumpfish living in the cold, cloudy, turgid water of the stream underneath the ridges and rough patches in the ice (Ekman 2010: 20).

Waiting in the life of Ekman's dog can be compared to the idea of waiting in Uexküll's famous tick. What awakens memory in the dog are the missing or expected sensory cues of sound or scent which could be turned into a desired and missing image as in the incident when 'he was awakened by a sound so familiar it called forth an image in him, bones cracking between powerful jaws' (Ekman 2010: 22). Now it is only the scent that is missing 'to complete the image, turning the crunching into his mother's teeth and the shadow down in the marsh into his mother' (ibid.). Even when the scent and the figure do not correspond to the image of his mother, 'his memory stretched and twisted to turn this long, thin back and the far too bristly tail into his mother' (ibid.). This encounter with the sounds of the mother is described in terms of 'vigilance and anticipation' and 'sheer longing' (ibid.) until it turns into terror at the scent of the fox. Interactions with 'the others' are first presented as part of his efforts at making sense of his new environment, connecting them to the familiar but now lost environment of house, mother, and humans. There are no species distinctions within the *Umwelt* of the dog but the narrator positions the human other and non-human others differently, thereby allowing for a linearity and teleology that move the circling narrative towards the end of the story.

Slowly, with the passing of time, 'the others' are present in the images formed in the forest. Paths, trails and scent constitute the forest environment: 'He never got really close to them but their paths criss-crossed his memory, the trails of their scents, their calls and chirps, the hoarse howl of the invisible one who was sometimes down by the lake' (Ekman 2010: 19).

> Sometimes there was another smell, dense, heavy and oppressive. It awakened memories but he had nothing to attach them to. He circled the holes drilled in the ice, sniffing around, pawing at the crust. He found wads of snuff and orange peel. His nose poked at them just as his memory did (Ekman 2010: 26).

The final return to human habitation is also made possible through an evocation of old memories combined with a sustained effort at creating new memories on the part of the man who finds the dog in the forest. Within these patterns of recurrent events of the novel as a whole, the memory of the dog is represented as a layer of significant acts, in the beginning more often in the form of reactions to mistaken stimuli but, with time, as he becomes 'deliberate and cautious' (Ekman 2010: 47), or 'skilled' (Ekman 2010: 62), there is the sense of development in the narrative that leads up to the closing in the form of a return to the beginning: 'a gentle stream of soft talking that awakened a strong urge in the young grey dog, in the midst of all his confusion' (Ekman 2010: 99). At this point in the novel there is also a brief shift in focalization, this time away from the dog to the unidentified man who arrives by boat to a cabin in the woods. He establishes a connection with the dog and, finally, through perseverance, turns the waiting of the dog, into a waiting for him: 'He would lie in the opening and wait. That was his place. He was the one who waited' (Ekman 2010: 122). The events are first described as experienced through the sensations of the dog:

> Between himself and the man something happened every time they met: the voice and the food. That was the good part. It was a warm stomach and a pleasant sensation sifting like strong sunlight through his fur. It touched nerves and awakened memories with no images (Ekman 2010: 116).

This return to human habitation as a closing event is, however, present throughout the novel, both as beginning and as an ending. The awareness of the mother dog and the awareness of humans remain and return, and these are also what make the reconnection possible.

> Everything that happens is inside him. It has already happened. Everything that happens is vivid within. He knows. It flares up, flashes like a wing in the dark night, settles again. It encompasses a life that has been lived.
>
> Remembering and forgetting are the same murky depths. Something swirls up from the sludge – he recognises it. It settles – he forgets but knows. He is just the hard mask over vivid things remembered, elusive things forgotten (Ekman 2010: 85).

This passage carries the tension expressed in Rilke's eighth Duino Elegy (1997 [1923]) between the present and the past, a possibility phrased as longing. This animal longing, in Rilke expressed as *Abstand* (distance) instead of *Atem* (breath), explores the opposition to the human as an observer. In Ekman's text the narrator also takes up a position from the outside and, while trying to describe the world as perceived by the dog, is essentially an on-looker. Whereas both Rilke and Uexküll, albeit in very different ways, use forms of address that include the reader, Ekman's narrator does not address or include the reader and imagines animal memory and longing as essentially remaining beyond human understanding. The closing section of the novel is therefore presented as a tale where the events experienced by the dog are rendered as a story told to a human audience:

> The man often told the story of how he'd started putting out food at the lakeshore and then in the boat. One day when the dog had become accustomed to standing in the boat he had carefully shoved it out. The dog had crouched down, ears pulled back (Ekman 2010: 130).

At this point in the narrative, the narrator shifts from the designation 'the dog' and 'him' to 'the grey dog' thereby indicating that he is now seen as one dog among many other dogs who need to be told apart. In the final section of the novel, the dog is also given a name: 'They called him Plucky. That name came to the man as he rowed back that Sunday morning when he had been allowed to touch him for the first time' (Ekman 2010: 129).[18] It is in the human touch and the human naming that the story of the dog comes to an end.

> The tale ends there. No one knows what he was listening for or what he had been through out there where no one had been able to see him. No one knows whether there is a word for whatever it is he's waiting for (Ekman 2010: 133).

In closing the story, the narrator thus firmly establishes the prerogative of a human presence watching, describing, and naming in order to tell a story of what the dog experiences. Through this return to the human, Ekman's novel gestures towards 'the open' while remaining within a tradition that stresses the human need for stories.

Notes

1 '*När börjar en händelse? Den börjar inte. Det var alltid något före*' (Ekman 1986: 5). '*Händelse*' refers more specifically to an event, to something that happens.
2 Kerstin Ekman was awarded the August (Strindberg) Prize for nonfiction in 2007 for *Masters of the Forest* (2008; originally *Herrarna i Skogen* [2007]) and the August Prize for fiction in 1993 for the novel *Blackwater* (1996).
3 Kerstin Ekman was elected member of the Swedish Academy in 1978 but has not attended meetings since 1989 in protest against the Academy's refusal to condemn the *fatwa* against Salman Rushdie.
4 See Wright (2000) for a discussion of Ekman's oeuvre up to the year 2000. Three book-length studies, Schottenius (1992), Andersdotter (2005), and Lindhé (2008), have been

published about Kerstin Ekman in Swedish and one, Ottesen (2009), in Danish. See Oscarson (2010), Paterson (2008a; 2008b), and Wright (1987; 1991; 1996; 2000) for critical work and interviews in English. Oscarson (2010), Paterson (2008a), Blikstad (2002), and Billing (2009) study her work from an ecological perspective.

5 *Klimatsorg* refers to both sorrow and mourning for the climate.

6 The translations from Swedish are mine unless otherwise indicated.

7 Ekman has a degree from Uppsala University where she studied Comparative Literature and German. Her novels have a deep intertextual relation with both earlier and contemporary literary texts. See Wright (1991; 2000) for a discussion of these intertextual elements in her fiction.

8 'Old forest makes its own time. It flourishes and dies back in a never-ending lifecycle that allows for both history and renewal. Religion, like wild nature, follows cyclical patterns and the central mystery is renewal. Now we reject both. Our preferred mode of thought is linear. Lines begin, and must end. The end is death' (Paterson 2008b: 45).

9 Ekman and Eriksson (2011: 16) refer to Linneaus and the expression *Nomina si pereunt, perit et cognitio rerum*: '*Förlorar du tingens namn så förlorar du också kunskapen om dem*' ['Without names, our knowledge of things would perish']. In their discussion of Linnaeus's *Iter Lapponicum* they do, however, point to the colonialist implications with processes of naming that built on the assumption that the Sami population encountered by the early botanists did not properly see flowers.

10 The Swedish romantic poet Erik Johan Stagnelius (1793–1823) was the son of a vicar on the island of Öland. *See Blomman* takes its title from one of his poems and in their analysis Ekman and Eriksson (2011) suggest that, despite the expressions of Romanticism, such as 'emerald-green', Stagnelius was inspired by the mosses and lichens of his early surroundings. The poem 'Näcken' was translated into English by Edmund Gosse and describes the encounter between a boy and the evil water spirit *Näcken* of Scandinavian folklore. The boy admonishes *Näcken* and tells him to stop playing his fiddle since he will never see the beauty of Paradise, the angels and the flowers of Eden. The *Näcken* of Stagnelius's poem has the freedom to roam fields and woods but will never be 'a child of God' (1972).

11 The original is as follows: '*Mitt förhållande till trollen hade jag när jag var liten och pappa berättade om dem. Det har räckt hela livet för en del trollskriverier. Pappas berättelser och senare Stagnelius dikt 'Näcken' har gett mig känslan för att det finns liv utan själ och att det kan finnas längtan efter att få en själ. Att livet utan vår religion, vårt samvete och alla våra intellektuella överbyggnader på hjärnan vore ödsligt*'.

12 For the development of Uexküll's concept of *Umwelt* in relation to the concept of *Milieu*, see Pollman (2013).

13 In addition to the work of Agamben (2004) and Calarco (2008), Uexküll's influence on posthumanism is also discussed by Buchanan (2008) and Parikka (2010).

14 Uexküll imagines differences in *Umwelten* also between different human beings. In *Umwelt und Innenwelt der Tiere* (1909) he mentions the artist Holbein as an example of a person who lived in a much richer world than other humans and who had the capacity to judge how much of his own vision others would be able to comprehend. See Winthrop-Young (2010) for a detailed discussion of Uexküll's ideas of human *Umwelt*.

15 This is Pollman's (2013) expression.

16 For a discussion of metalepsis, see, for instance, Genette (1980: 234–5), for ontological metalepsis, see Bell and Alber (2012).

17 See, for instance, David Herman's *Story Logic* (2002: 280–2) for a discussion of spatialization and the distinction between topological and projective locations.

18 In the original, the dog is named 'Tapper' (Brave), a name often given to soldiers with a much more serious ring to it than the name 'Plucky'. The naming of the dog is also the result of a deliberate thinking process rather than a name that simply 'came to him'.

References

Agamben, G., 2004. *The open: man and animal*. Trans. Kevin Attell. Stanford, US: Stanford University Press.

Andersdotter, A., 2005. *Det mörka våldet: spåren av en subjektsprocess i Kerstin Ekmans författarskap*. Stockholm: Brutus Östlings Bokförlag Symposion.

Bell, A. and Alber, J., 2012. Ontological metalepsis and unnatural narratology. *Journal of Narrative Theory*, 42 (2), 166–92.

Billing, A., 2009. Jord och tid: en ekokritisk läsning av Kerstin Ekmans Häxringarna. *In:* S. Packalén, ed. *Ekokritik, Jean-Henri Fabre, återvinningens estetik, Kerstin Ekman, posthumanism, elektriska får och mekaniska människor, djurkaraktärer, kultur och hållbar utveckling*. Västerås, Sweden: Mälardalens Högskola, 58–73.

Blikstad, L., 2002. Intervju med Kerstin Ekman. *In:* S. Farran-Lee, ed. *.doc. Manus*. Stockholm: Norstedt, 63–82.

Buchanan, B., 2008. *Onto-ethologies: The animal environments of Uexküll, Heidegger, Merleau-Ponty, and Deleuze*. Albany, US: State University of New York Press.

Calarco, M., 2008. *The question of the animal from Heidegger to Derrida*. New York: Columbia University Press.

Genette, G., 1980. *Narrative discourse: an essay in method*. Trans. Jane Lewin. Ithaca, US: Cornell University Press.

Ekman, K., 1986. *Hunden*. Stockholm: Albert Bonniers Förlag.

Ekman, K., 1988. *Rövarna i Skuleskogen. Roman*. Stockholm: Albert Bonniers Förlag.

Ekman, K., 1993. *Händelser vid vatten*. Stockholm: Albert Bonniers Förlag.

Ekman, K., 1996. *Blackwater*. Trans. Joan Tate. London: Vintage.

Ekman, K., 1999. *Forest of hours*. Trans. Anna Paterson. London: Random House.

Ekman, K., 2007. *Herrarna i skogen*. Stockholm: Albert Bonniers Förlag.

Ekman, K., 2008. Selections from *Masters of the forest*. Trans. Anna Paterson. *World Literature Today*, 82 (4), 47–9.

Ekman, K., 2010 [1996]. *The dog*. Trans. Linda Schenk and Rochelle Wright. London: Sphere.

Ekman, K. and Eriksson, G., 2011. *Se blomman*. Stockholm: Albert Bonniers Förlag.

Faulkner, J., 2011. Negotiating vulnerability through 'animal' and 'child': Agamben and Ranciere at the limit of being human. *Angelaki: Journal of the Theoretical Humanities*, 16 (4), 73–85.

Genette, G., 1980. *Narrative discourse: an essay in method*. Ithaca, US: Cornell University Press.

Haraway, D., 2008. *When species meet*. Minneapolis, US: University of Minnesota Press.

Heidegger, M., 1995. *The fundamental concepts of metaphysics: world, finitude, solitude*. Trans. William McNeill and Nicholas Walker. Bloomington, US: Indiana University Press.

Herman, D., 2002. *Story logic: problems and possibilities of narrative*. Lincoln, US: University of Nebraska Press.

Herman, D., 2009. *Basic elements of narrative*. Chichester, UK: Wiley-Blackwell.

Jörneberg, J., 2013. Att kliva av från asfalten och se: en intervju med Kerstin Ekman. *Tidskriften 10TAL*, 12/13, December, 38–43.

Lindhé, C., 2008. *Visuella vändningar: bild och estetik i Kerstin Ekmans författarskap*. Uppsala, Sweden: Svenska Litteratursällskapet.

McHugh, S., 2011. *Animal stories: narrating across species lines*. Minneapolis, US: University of Minnesota Press.

Oscarson, C., 2010. Where the ground answers the foot: Kerstin Ekman, ecology, and the sense of place in a globalized world. *Ecozon@ – European Journal of Literature, Culture and Environment*, 1 (2), 8–21.

Ottesen, D., 2009. *Barmhjertighed: en fortælling om Kerstin Ekman og hendes forfatterskap.* Copenhagen: Gyldendal.

Parikka, J., 2010. *Insect media: an archaeology of animals and technology.* Minneapolis, US: University of Minnesota Press.

Paterson, A., 2008a. Landscapes remembered: Kerstin Ekman and nature. *World Literature Today,* 82 (4), 40–2.

Paterson, A., 2008b. Mistress of the forest: an interview with Kerstin Ekman. *World Literature Today,* 82 (4), 43–6.

Pollman, I., 2013. Invisible worlds, visible: Uexküll's Umwelt, film, and film theory. *Critical Inquiry,* 39 (4), 214–38.

Rilke, R.M., 1997. Die achte Elegie. *In:* R.M. Rilke, *Duineser Elegien: Gedichte.* Stuttgart: Reclam, 200–2.

Schottenius, M., 1992. *Den kvinnliga hemligheten: en studie i Kerstin Ekmans berättarkonst.* Stockholm: Albert Bonniers Förlag.

Stagnelius, E.J., 1972. Näcken. *In:* E.J. Stagnelius, *Dikter.* Stockholm: Albert Bonniers Förlag, 129.

Uexküll, J. von, 1909. *Umwelt und Innenwelt der Tiere.* Berlin: Verlag von Julius Springer.

Uexküll, J. von, 2010. *A foray into the worlds of animals and humans: with a theory of meaning.* Trans. Joseph D. O'Neil. Minneapolis, US: University of Minnesota Press.

Winthrop-Young, G., 2010. Afterword: Bubbles and webs: a backdoor stroll through the readings of Uexküll. *In:* J. von Uexküll *A foray into the worlds of animals and humans: with a theory of meaning.* Trans. Joseph D. O'Neil. Minneapolis, US: University of Minnesota Press, 209–43.

Wright, R., 1987. Theme, image, and narrative perspective in Kerstin Ekman's *En Stad av ljus. Scandinavian Studies,* 1, 1–27.

Wright, R., 1991. Approaches to history in the works of Kerstin Ekman. *Scandinavian Studies,* 63 (3), 293–305.

Wright, R., 1996. Androgyny in Kerstin Ekman's *Rövarna i Skuleskogen* and Virginia Woolf's *Orlando. In:* H. Kress, ed. *Litteratur og kjønn i Norden.* Reykjavík: Háskólaútgáfan, 676–81.

Wright, R., 2000. Textual dialogue and the humanistic tradition: Kerstin Ekman's *Gör mig levande igen. Scandinavian Studies,* 3, 279–300.

Ziarek, K., 2008. After humanism: Agamben and Heidegger. *South Atlantic Quarterly,* 107 (1), 187–209.

8 Transcultural affect
Human–horse relations in Joe Johnston's *Hidalgo*, Steven Spielberg's *War Horse*, and Belá Tarr's *The Turin Horse*

Sissy Helff

In this chapter I seek to explore both the visual poetics and the cultural politics of transcultural affect in recent European and Hollywood cinema dealing with the representation of human–horse relations. Covering a timespan of approximately thirty years, namely the period between 1889 and 1919, all of my selected films scrutinize the relationships between humans and horses in modern times. Generally speaking, the modern period was a period of great social and cultural unrest and change; on the one hand, it was a period marked by massacres of the Indian population in the Americas while, on the other, it was a time characterized by a growing feeling of insecurity in Europe. In a nutshell, it was a time of contrasts. Despite the fact that the invention of the first movie camera has framed our perception of the world in a hitherto unknown way, notions of humiliation and disillusionment linked to the Great War tested people's faith in the positive power of technological progress. After all, it was in these years too that Freud, the father of psychoanalysis, introduced the unconscious as a central category for exploring emotional wastelands and desires. All of these topics will be approached through the selected movies and my analysis of the films. By approaching the relationship between humans and horses and the underlying politics of transcultural affect in a comparative framework, I will shed light on a visual poetics that is informed by the previously mentioned historical incidents as much as it draws on individual negotiations of entangled modernities: memory, mourning, melancholia, trauma, and guilt. Examples will be drawn from films as diverse as the black-and-white auteur movie *The Turin Horse* (2011), by Béla Tarr, the war movie *War Horse* (2011), by Steven Spielberg, and Joe Johnston's sports-adventure-western *Hidalgo* (2004). While all three movies focus on human–horse relations, they engage differently with compassion, feeling, affect, and emotion.

As early as at the beginning of the twentieth century the Russian film maker-cum-theorist Sergei Eisenstein championed emotions as a key concern in cinema. In his early texts (1988: 39) he described the power of the 'cinema as a factor for exercising emotional influence over the masses'. This observation characterizes all of Eisenstein's theoretical and filmic works thus '[w]ithout the preconditions of love and hatred', Eisenstein (1991: 294) argues, 'no work of art can come into being'. While Eisenstein emphasized the importance of the contemporary conglomeration of emotions, movement, and film making, more than half a century had to go by until cultural and literary theorists re-discovered affect and emotional registers as analytical categories in their deductions of our world of high modernity. It was in the mid-1990s that in the social sciences and humanities a general turn towards affect with a particular interest in the neurosciences of emotion was postulated (see, e.g., Reddy 2001). In consequence, a number of literary and cultural scientists called for the renewal of their disciplines and their respective methodological toolkits. The cultural studies scholar Eric Shouse (2005: para. 12) stated that 'the importance of affect rests upon the fact that in many cases the message consciously received may be of less import to the receiver of that message than his or her non-conscious affective resonances with the·source of the message'. Accordingly, Shouse (2005: para. 14) concluded that the influence of media formats therefore resides 'not so much in their ideological effects, but in their ability to create affective resonances independent of content or meaning'. Shouse's approach echoes in many ways how Brian Massumi (2002: 28) understands affect, namely, as non-cognitive, physical processes which are 'irreducibly bodily and autonomic'. With these assumptions, Shouse and Massumi follow the ideas of the American psychologist-cum-philosopher Silvan Tomkins, who in the 1940s formulated his theory of affect. Based on a typology of nine distinct affects (Tomkins 1995: 64–100), Tomkins's approach suggests, as Ruth Leys (2011: 437) writes, 'that the affects are only contingently related to objects in the world' and that 'our basic emotions operate blindly because they have no inherent knowledge of, or relation to, the objects or situations that trigger them'. This tension between affect, feeling, and emotion on the one hand, and cognition on the other, characterizes discussions around an ostensibly incompatible condition.

In a recent article the psychological constructivists William A. Cunningham, Kristen A. Dunfield, and Paul E. Stillman (2013: 345) add a fresh perspective as they argue 'that the distinction between "emotion" and "cognition" is a false dichotomy'. By applying a broader view to cognition (so that it encompasses any information processing) the three psychologists conclude that emotion

> is inseparable from cognition, in that all mental operations require some form of information processing. Indeed, when considering the subjective experience of emotion, it is probable that the underlying cognitive processes are complex and multifaceted, and possible that when more fully articulated may not even correspond to current linguistic categories (Cunningham *et al.* 2013: 345).

In *The Affective Turn: Theorizing the Social*, the cultural theorists Patricia Clough and Jean Halley (2007) have argued that being attentive to the affective turn always includes theorizing the social. To Clough (2007: 3) 'affect' reflects on our social form of being in the world. Seen through such a lens, cultural studies dealing with affect are interdisciplinary and political by definition. At this point in my study it might be helpful to introduce working definitions for *feeling*, *emotion*, and *affect*, notwithstanding the many complications which go hand in hand with an attempt to define these words and categories with their underlying concepts.

For this reason, I will neither introduce nor work with a catalogue of different affect and emotion programmes, as was once proposed by Silvan Tomkins; rather, I seek to illuminate working definitions which may be more precise for dealing with affect and emotional registers. In the editorial of a special issue on affect and feeling Lisa Blackman and John Cromby (2007: 5–6; emphasis original) offer a quite useful definition of the related concepts:

> where *feeling* is often used to refer to phenomenological or subjective experiences, *affect* is often taken to refer to a force or intensity that can belie the movement of the subject who is always in a process of becoming. […] *Affects* do not refer to a 'thing' or substance, but rather the processes that produce bodies as always open to others, human and non-human, and as unfinished rather than stable entities. […] *Emotions*, in contrast, are those patterned brain/ body responses that are culturally recognisable and provide some unity, stability and coherence to the felt dimensions of our relational encounters.

By understanding affect as form of emotional contact in the various visualizations of horse–human relationships, my analyses of my selected feature films will illustrate that a distinctively transcultural conceptualization of affect is needed in order to decode the pictures and to write more sufficiently about affect in film. One major concern is to read affect as an integral part of our global world, a world formed by multiple and at times even entangled modernities. In this respect, I find Göran Therborn's conceptualization of entangled modernities particularly helpful since, in line with Shalini Randeria's (1999: 373–82) conceptualization of entangled modernities, he emphasizes 'not just the co-existence of different modernities but also their interrelations, current as well as historical' (Therborn 2003: 295). Thus when talking about the filmic representation of the horse–human relationship I suggest thinking inter-relationally about affect. Thus, my definition of transcultural affect is rooted in a modernity discourse which sets out to provincialize Europe by suggesting not a mainly European intellectual heritage, but a global one (Chakrabarty 2000: 4). It emphasizes the need to incorporate a decentralized thinking about one's being in the world with an understanding of the multiple creative practices which come alive in, and are visualized through, film and cinema. Such a politico-aesthetic thinking recognizes how and why political realities are differently reconfigured and mediated in various regions in the world at several interrelated moments in time.

Visualizing the relationship between humans and horses in film

While in the two Hollywood productions by Spielberg and Johnston the story forms the centre of the movies, its role in Béla Tarr's auteur film is of less significance, as the Hungarian film maker states in an interview:

> I don't care about stories. I never did. Every story is the same. We have no new stories. We're just repeating the same ones. I really don't think, when you do

a movie, that you have to think about the story. The film isn't the story. It's mostly picture, sound, a lot of emotions. The stories are just covering something […] if you're a Hollywood studio professional, you could tell [the] story in 20 minutes. It's simple. Why did I take so long? Because I didn't want to show you the story. I wanted to show […] life (Kohn 2012).

So what then is Béla Tarr's *The Turin Horse* about? At the Berlin Biennale film festival, Tarr introduced his film as follows:

In Turin on 3rd January, 1889, Friedrich Nietzsche steps out of the doorway of number six, Via Carlo Alberto. Not far from him, the driver of a hansom cab is having trouble with a stubborn horse. Despite all his urging, the horse refuses to move, whereupon the driver loses his patience and takes his whip to it. Nietzsche comes up to the throng and puts an end to the brutal scene, throwing his arms around the horse's neck, sobbing. His landlord takes him home, he lies motionless and silent for two days on a divan until he mutters the obligatory last words, and lives for another ten years, silent and demented, cared for by his mother and sisters. We do not know what happened to the horse (Berlinale 2011).

The Turin Horse starts at this moment; in its course, the slow-burn narrative reflects on the minimal conditions of survival, telling us about the sombre lives of the driver, his adopted daughter, and their nameless horse. The movie's slender story line unfurls, and the gloomy situation grows even darker after the horse refuses to participate in the daily work and eventually even stops eating. The farm's only water source, a well, starts to dry up and the situation descends into an overarching bleakness, which is visualized in the fading light of the kerosene lamp. The film's minimalist long takes and the black-and-white photography exist virtually without dialogue and remind us of silent film. Tarr's anti-narrative thus throws us back to a much earlier form of cinema while, moving fast-forward, introducing a new cinematic experience. This effect emerges from the combination and interconnection of various cinematic styles, moments of film making, and cinematic history, thus re-narrating the becoming of cinema as the result of an entangled European modernity. It is at this moment of transcultural becoming that the director causes us to experience his own affects involved in the process of filmmaking.

On several occasions the director has pointed out that in his films a storyline is not only information and cut, but that it works with a particular sense of cinematic space in combination with an effective utilization of movie time. *The Turin Horse* opens with a black screen, an 'empty' picture pregnant with meaning, signalling the first and the last picture of the film, or even the director's oeuvre, while the backstory is briefly told in Hungarian from the start. This opening sequence already demonstrates Tarr's complex composition since the spoken language used throughout does not correspond with the location of the plot and the images that the film endeavours to convey. The emerging irritation expresses a sense of deterritorialization and unbelonging, feelings which grow not only within the film's own story world, but also within its viewers. This cinematic technique contributes to a creative entangling of histories and memories of hitherto untold stories of European rural modernities on the one hand, while on the other hand portraying the auteur's own emotions of dissatisfaction and fear concerning the future of filmmaking. Following this train of thought, one may also wish to read such cinematic practices as visualizations of transnational memories of Béla Tarr's own transcultural affect concerning the cinematic tradition of storytelling.

Yet as soon as this short introductory note is told and the Hungarian voice-over falls silent, the black screen is ready to hand over to the camera, which immediately introduces the main character, the horse, from multiple camera angles and movements, as the non-human animal is walking along a dusty trail across broad acres of open land and through the woods. When the camera occasionally adopts a worm's-eye view of the horse, viewers may even feel that they are being run over by the cart-pulling creature. In a way, this scene works as a mirror to the above-mentioned black screen, since the horse's presence is accompanied by soundscape dominated by string instruments that swallow up of all of the horse sounds, including the clattering horseshoes. It is in this emotionally dense cinematic passage that the audience experiences a physical and emotional closeness to the horse. Tarr's dramatization of the horse clearly places the non-human animal at the film's centre and makes all of the other actors appear marginal. Nevertheless, as some critics have noted, especially the farmer and his daughter radiate a particular kind of sacredness that mirrors the dignity of the horse. This first horse-centred walking sequence eventually dissolves in a cinematic rapprochement between the world of the non-human animal and the human beings when the string instrument tune gives way to the sounds of the horse and cart. The camera movement, the sound, and the celluloid texture of the movie generate a unique watching experience since these elements connect the audience, the driver, and his daughter both physically and emotionally with the horse.

Looking more deeply into the film's narrative structure, we learn that the film is organized according to the passing days. At the end of the first day, the father, old Ohlsdorfer, remarks that for the first time in 58 years he is unable to hear the woodworms working; and when a little later the daughter asks him: 'What is it all about?' and the father replies 'I don't know', 'Let's sleep' [28:33–30:29], it has become clear that change is about to come. Whereas the human beings are still wondering about the potential consequences, it seems as if the horse has already made up its mind. Thus, from the second day onwards the horse stops participating in the farmer's daily work, no matter how hard Ohlsdorfer, who is portrayed through his paralysis, tries to drive the non-human animal onwards. The irrevocable steering towards change is already indicated in this initial scene, but the motif also, in fact, organizes the whole film and hence is repeated in a number of subplots, for example, in the neighbour's visit or in the passing by of the Roma clan. All such encounters emphasize feelings of uneasiness and emotions of fear, and also in broader terms the film's own negotiation of cultural change. While Ohlsdorfer refuses to see the destruction of his home and *Lebenswelt* at the beginning, it is his loyal working animal that gains autonomy and agency by refusing first to work and later to eat. This action for a cause may be described as 'minimal rationality' (Dretske 2006: 107–16), since the horse takes actions for a specific reason. Revisiting our previous definition of affect, namely that 'affect is often taken to refer to a force or intensity that can belie the movement of the subject who is always in a process of becoming' (Blackman and Cromby 2007: 5), it becomes clear that the horse's agency reformulates its relationship to the Ohlsdorfers, an action which affects the lives of both the farmer and his daughter. With this conscious act of refusal the faithful horse, like the auteur himself, adopts a stance and physically comments on, and possibly also silently critiques, its being in the world and the hostile environment that it has to endure.

Questions about the chosen emotional approaches to the making, screening, and watching of films are central for acquiring a better understanding of Béla Tarr's *The Turin Horse* and the movie's reception. In an interview Tarr replies to the question of whether his movie is based on an apocryphal story about Nietzsche feeling sympathy for a beaten horse in the following way: '[l]isten to your heart and trust your eyes. That's enough' (Kohn 2012). Like Eisenstein, Tarr strictly believes in the power of cinema to unlock emotions. He is fully aware of the responsibility he carries as a filmmaker and of the authority which accompanies his filmmaking, when he states: '[w]hen I am doing a film, I know the whole thing before I shoot it. Really, I have a very strong vision for all my movies. I know very clearly how and what I'll do' (ibid.). If we take Tarr at his word, it is significant to note that *The Turin Horse* is likely to have been his final film and, as such, not only does it close an auteur's filmic oeuvre, but also the movie itself assesses Europe's high modernity and hidden nationalistic fundamentalisms. Hence, the movie is, in addition, critiquing the current conservative Hungarian establishment, which has been systematically stifling its country's creative film scene. Despite the relentless warnings concerning the diminishing Hungarian cinema issued by Tarr and others, the first effects of this reactionary cultural politics can be seen in the drastic cutting of national funding in recent years, ending Hungary's national film festival 'The Week of Hungarian Film' in 2012, scarcely a year after the light from Ohlsdorfer's kerosene lamp faded in Béla Tarr's *Turin Horse*, released in 2011.

According to Steven Spielberg, his feature film *War Horse* (2011), a US and British coproduction, is anything but a war movie. Based on Michael Morpurgo's (2010 [1982]) successful novel *War Horse* and an enormously popular international theatrical production,[1] Spielberg's filmic adaptation, although referring to combat in its title, is, according to Spielberg, mainly an epic adventure story for audiences of all ages (BBC News, Entertainment and Arts 2012). Set against a sweeping canvas of rural England and continental Europe during the Great War, in a traditional Hollywood horse-film manner the Hollywood movie sets out to narrate a friendship between a foal named Joey and Albert, a teenage farmer's boy, who cares for and trains the horse. As a result of their difficult living conditions, however, Albert's family soon has to sell the horse, and this initiates Joey's journey through the Great War, including different countries and ownerships – the British cavalry, German soldiers, and a French farmer and his granddaughter. The movie's classic horse story is transformed into a war horse movie whose filmic climax is reached on the battlefield, namely when, at the heart of the disputed 'no man's land', Joey, the war horse, is badly wounded following his panic-induced attempt to break through barricades of barbed wire securing the trenches of the opposing armies. It is in this very cinematic moment that the war in the trenches comes to a halt, soldiers suddenly appear as human beings, and the opposing parties even decide to stop their gun battle until the horse is freed from the barbed wire shackles and taken away from the battlefield. The severely wounded horse, however, has given up on life. He stops eating and shows almost no reactions and affect towards the caring soldiers. Thus, in an act of mercy an officer sets out to shoot the more dead than alive non-human animal. However, it is then that the horse recognizes his friend Albert's distant whistling and starts responding again. Joey's memory of an intact earlier emotional bond, his friendship with Albert, brings him back to life. While this scene of friendship certainly does not shy away from flirting with sentimental kitsch, it nonetheless also conveys feelings of disillusionment concerning technical progress, in this case of modern warfare, and a universal yearning for a life beyond war, both emotions that were quite prominent in the years following the Great War.

A remarkable aspect of Spielberg's *War Horse* is the director's masterful visualization of an equation that interrelates the wartime experiences of a non-human animal with the experiences of human beings. In this equation the non-human animal becomes an icon of the wounded and a further victim of the Great War. Shared by the million war horses sent to the front line in the conflict, Joey's fate and his story stand in the many untold narratives of the Great War (see also Butler 2011). While we all are aware of the horrifying iconic images of the wounded and dying soldiers in the trenches during the Great War, the sequences revealing the exhausted and emaciated horses on the war's front line push the watchers through an extreme viewing experience: in these shots the cinemagoer is confronted with patterns of violence and cruelty which are generically 'hostile' to classic horse films. The visualization of both the cruelty of the Great War and the overall hopelessness surrounding both human beings and the dependent non-human animals caught in inhumane circumstances acts as a shapeshifting moment that turns the horse film into a war movie. Thus, Spielberg's *War Horse* visually translates the complex transcultural history of the Great War into a mnemonic space of transcultural affect; affect is here meant to describe both, the emergence of a new cinematic genre which masterfully combines emotions usually connected to the horse film with emotions commonly related to war movies, as well as a new cinematic experience which informs and affects our overall viewing pattern.

Ultimately, the film has radically changed my perspective on what constitutes the genre of the horse film, since *War Horse* combines ostensibly incompatible, visual viewing patterns. With great effectiveness, at the end of the film Spielberg dares to reconnect the final sequence visually with the opening scenes and the earlier horse-film-blockbuster-charm when the two friends, Albert and Joey, are finally reunited and the audience shares their emotional homecoming to England and the family's simple farm life. Kitsch-like or not, the whole sequence has been filmed through an orange-tinted camera lens which allows the beams of warm, afternoon sunlight shine into every cinemagoer's heart. As the audience watches the two friends trotting off into a peaceful afternoon's red sunset and, eventually, even through the family farm's gate, the romantic transfiguration has done its trick again by offering a placatory gesture which transforms the war horse film back into a feel-good horse movie for all generations.

Like Spielberg's *War Horse*, Joe Johnston's sport film-cum-adventure western *Hidalgo* (2004) invests to a major extent in the mixing of filmic genres, taking its viewers on a journey to explore various emotional registers. The Disney film story is set in North America, probably somewhere in South Dakota, immediately prior to the Wounded Knee Massacre, i.e., at some point in December 1890. It is then that we first meet the film's hero, the American Frank Hopkins, and his mustang Hidalgo, while the two are participating in a cross-country race, which they ultimately win. The result of the race is not too well received by their opponents, and serves as a cliffhanger that is used to introduce the main motif of the movie, racism. While initially the casino scene appears to occur quite innocently, it is here that the movie introduces Hopkins as a man who stands up for his 'people'. Thus, the cowboy knocks his former competitor down after the latter insults Hidalgo by challenging the horse's overall right to participate in races, given that 'mixed-bloods' should not be allowed to race against thoroughbreds (see also Kollin 2010: 18). While Hopkins can easily resolve this first fight for himself, this conflict only foreshadows a much more terrifying battle. Thus, the motif of racism is further enforced through sequences depicting Hopkins witnessing the Wounded Knee Massacre and the fate of the Sioux during and after this catastrophe. The tragedy not only cost the Native Americans their homes and grassland but also caused the decline of the mustang herds, since these non-human animals did not fit into the conception held by some of what North America should become.

The film's visual language is controlled by a number of iconic images, of which the powerful, historical photograph of the Miniconjou chief Spotted Elk, more popularly known as Bigfoot, is particularly striking. The iconography of exploitation and slaughter is unmistakably evoked in the movie's massacre sequence depicting the killed chief in a position that is similar to that in the historical source – an old, unarmed man lying on the ground, covered with white ashes and holding a white flag in his right hand. While the flag is an additional statement by the film director, since it is missing in the photograph, it accentuates the movie's intention to 'screen back' at – that is, in a sense, to scream back at – racism and genocide. At this point in the movie the audience is unaware that Hopkins himself is the son of a Sioux woman. What the audience does immediately realize is that these experiences destroy the man's life. The cowboy leaves the carnage deeply traumatized; he cannot live with his haunting memory and his feeling of guilt induced by not having stood up for his people. This provides an explanation for his becoming a drunkard, but also an actor who earns his living by constantly reenacting his former cowboy self and the massacre of the Wounded Knee at county fair spectacles. In one of these performances the cowboy and his horse are seen by an Arabian sheikh, who invites the broken man to participate in an ancient tradition, the Ocean of Fire race, a 3,000-mile survival contest staged across the Arabian Desert. Hopkins takes this opportunity and from this moment onwards, the western becomes an adventure sports film. It is the adventure that helps him by overcoming his past. With the advertised trophy money in mind and his friend Hidalgo at his side, Hopkins attempts to make the impossible possible. So it is no wonder that the ambitious combination of genres and storylines almost collapses under the movie's own multifaceted story world. The film follows the saying that when there is no way forward, adventure helps – and to some extent this remains true. Nonetheless, it would not be a Disney production if the heroes did not finally succeed in not only surviving the desert, but also in winning the race. Accordingly, *Hidalgo* closes on sequences that narrate the comrades' successful homecoming: with the prize money, Hopkins is able to buy back some of the wild mustangs. Together with Hidalgo the cowboy sets the herd free; and like Hidalgo the cowboy Hopkins also decides to leave behind his former profession in order to live with his tribe. In sum, the emotional registers utilized in the movie are not really new, and in comparison with the other movies discussed here, *Hidalgo* is the most conventional of the films in terms of its visual language, sound, and storytelling. The movie nevertheless convincingly introduces a serious discussion about race and racism. Thus, while nobody can turn back time to alter the series of events, Hopkins's redemption of the horses is an act of reparation that offers some relief not only to the cowboy and his horse but equally so to the film's watchers. It is the emotionally dense horseman–horse relationship and the complex cowboy-figure that create the visual frame of the movie and prevent the visual narrative from falling apart.

While some may wish to explain the movie's exceptional story in terms of the film representing a larger-than-life experience, and since the film claims to retell a 'true-life story', others are more sceptical about this endeavour. Thus, it is hardly surprising that, following the film's release, it provoked a huge discussion concerning the extent to which the movie presents a true or less-than-true life story. A number of critics and journalists clearly challenge the creative freedom utilized by the scriptwriter, John Fusco. In addition, they criticize Disney's insistence on selling the movie as a 'true story' retelling an extract of North American history (see O'Reilly and O'Reilly 2004; Desai 2004; Kollin 2010). While neither Fusco nor the film studio considered it necessary to read the official historical correspondence between the rulers of Hail, the Sublime Porte in Istanbul, and the Khedive in Cairo, the writer Anuj Desai has pointed out that these papers might have served as ideal materials for verifying or clarifying Hopkins's life story, since the film has been marketed on the basis of the studio's 'truth' claims. If the script writer and the studio had decided to investigate the historical correspondence, it would have soon become clear that the race supposedly mentioned in these papers is indeed present only in its absence. 'Not a single word about this race', writes Desai (2004), quoting several critics. In line with the researchers, she points to the tendency to exoticize the Middle East to such an extent that fact and fiction merge, since no matter how hard 'Westerners – Disney and Fusco among them – would like to believe that Arabs relied exclusively on an oral tradition "the literature of the horse extends back to before the coming of Islam"' (ibid.; see also Kollin 2010: 20–1). No matter whether Hidalgo's story is true or simply a tall tale, what I consider most surprising is the great media stir that the story itself produced and the extent to which people seem to have been affected by the question of truth or lie (see Long Riders' Guild 2014). Perhaps it is this discussion that turns a formerly rather affectless movie into a media effect that affected many people around the world.

Conclusion

Given that affect, as a form of emotional contact, is central in the visualization of horse–human relationships in contemporary film, my analysis introduces and draws on a distinctively transcultural conceptualization of affect. One major concern is to read affect as an integral part of our global world, a world formed by multiple and at times even entangled modernities that not only coexist but also come alive in their current and historical interrelations (see also Therborn 2003: 295). Hence, when talking about the representation of the horse–human relationships in film, I would suggest that it is necessary to think inter-relationally about affect. My definition of transcultural affect is thus rooted in a modernity discourse that sets out to provincialize Europe by suggesting, not a mainly European cinematic intellectual heritage, but a global one. Following this train of thought, my decentralized thinking includes the artistic practices of the film directors and actors, and also the watching habits of the cinemagoers, as they come alive in each new screening of a particular film. As a result, each film screening foregrounds a unique take on our being in the world. Such politico-aesthetic thinking recognizes how and why political realities are differently reconfigured and mediated in various parts of the world, at several interrelated moments in time.

The films selected for this study present a panorama of horse–human relations that unfold against a backdrop of global, historical moments (Nietzsche's nervous breakdown, WWI, the Wounded Knee Massacre) and the relevant historical sites (Turin, WWI battlefields, the Wounded Knee Massacre site). All films exemplify the film makers' strong faith in the power of cinema to unlock emotions. Yet throughout Tarr's movie, the audience feels that the director is fully aware of his own responsibility as a filmmaker. Nothing appears accidental. For Tarr, filmic form and sound are central to the task of addressing the audience's emotions. Hence, *Turin Horse*, which is presumably his final movie, participates in a configuration of Europe's high modernity on the verge of decay. The film formulates its critique not simply at the level of plot but even more so in its aesthetics. Tarr's critique is all about ideology, in particular, the continuously burning candle of nationalistic fundamentalisms. Spielberg's *War Horse*, in contrast, conveys a considerably lighter message. The film translates the complex transcultural history of the Great War visually into a mnemonic space of high modernity and transcultural affect – affect is here meant to describe both – signalling a change in perspective about an iconic moment in the past as well as the emergence of a new cinematic genre that masterfully combines the emotions usually related to the horse film with emotions commonly related to war movies. By generating this new transgeneric cinematic experience, Spielberg affects our overall viewing pattern. Joe Johnston's *Hidalgo* also invests in the mixing of filmic genres in order to reach out emotionally to the cinemagoer. While the newly created genre of the sport film-cum-adventure western may sound as attractive as it is challenging with respect to keeping its narrative coherency intact, the biggest challenge is certainly reflected in the discussion dealing with the 'real story' behind the film. It is here that audiences are taken on journeys in which they may explore emotional registers that contest our common perception of fact and fiction as well as our common viewing habits.

As these three films exemplify, affect, as a form of emotional contact, has always been, and still is, central in the visual narration of horse–human relationships in contemporary cinema. The concept of transcultural affect seeks to emphasize the importance of understanding film as a contextualized art form rooted in local production set-ups with transnational links while employing images and sounds ingrained in transcultural aesthetics. Transcultural affect, thus, is part of an entangled modernity which advocates a global intellectual heritage while emphasizing a politico-aesthetic thinking that recognizes how and why political realities are differently reconfigured, rememorized, and mediated around the world at several interrelated moments in time.

Note

1 The performance *War Horse* was a 2007 theatrical production of the National Theatre of Great Britain using full-sized horse puppets performed by the South African Handspring Puppet Company to bring breathing, galloping, full-scale horses to life on the stage. The performance has been staged in five countries including the UK, Ireland, Australia, North America, and Germany, and was broadcast live from London's West End to cinemas around the world on 27 February 2014.

References

BBC News, Entertainment and Arts, 2012. An interview with Steven Spielberg: '*War Horse* is not a war movie' [online]. 9 January 2012. Available from; www.bbc.co.uk/news/entertainment-arts-16466772 [Accessed 2 March 2015].

Berlinale, 2011. Film file: *A torinói ló, The Turin Horse, Das turiner Pferd* [online]. Available from: www.berlinale.de/en/archiv/jahresarchive/2011/02_programm_2011/02_Filmdatenblatt_2011_20115947.php#tab=filmStills [Accessed 3 March 2015].

Blackman, L. and Cromby, J., 2007. Editorial: affect and feeling. *International Journal of Critical Psychology*, 21, 5–22.

Butler, S., 2011. *The war horses: the tragic fate of a million horses in the First World War.* Wellington, UK: Halsgrove.

Chakrabarty, D., 2000. *Provincializing Europe*. Princeton, US: Princeton University Press.

Clough, P.T., 2007. Introduction. *In*: P. Clough with J. Halley, eds. *The affective turn: theorizing the social*. Durham, US: Duke University Press, 1–33.

Clough, P.T. with J. Halley, eds., 2007. *The affective turn: theorizing the social*. Durham, US: Duke University Press.

Cunningham, W., Dunfield, K., and Stillman, P., 2013. Emotional states from affective dynamics. *Emotion Review*, 5 (4), 344–55.

Desai, A., 2004. A mirage in the desert: Viggo Mortensen's *Hidalgo* is based on a not-so-true story. *Slate Magazine* [online], 4 March. Available from: www.slate.com/articles/news_ and_politics/life_and_art/2004/03/a_mirage_in_the_desert.html [Accessed 4 March 2015].

Dretske, F., 2006. Minimal rationality in rational animals? *In*: S. Hurley and M. Nudds, eds. *Rational animals?* Oxford, UK: Oxford University Press, 107–16.

Eisenstein, S., 1988. *Writings, 1922–1934*. Trans. and ed. R. Taylor. Bloomington, US: Indiana University Press.

Eisenstein, S., 1991. *Towards a theory of montage*. Trans. M. Glenny. M. Glenny and R. Taylor, eds. London: BFI Publishing.

Hidalgo, 2004. Directed by Joe Johnston. USA: Touchstone Pictures.

Kohn, E., 2012. An interview with Bela Tarr: why he says 'The Turin Horse' is his final film. *Indiewire* [online], 9 February. Available from: www.indiewire.com/article/bela-tarr-explains-why-the-turin-horse-is-his-final-film [Accessed 3 March 2015].

Kollin, S., 2010. 'Remember, you're the good guy': *Hidalgo*, American identity, and histories of the western. *American Studies*, 51 (1/2), 5–25.

Leys, R., 2011. The turn to affect: a critique. *Critical Inquiry*, 37 (3), 434–72.

Long Riders' Guild, 2014. Walt Disney and Hidalgo – A decade of deceit 2014 [online]. Available from: www.thelongridersguild.com/hopkins2014.htm [Accessed 3 March 2015].

Massumi, B., 2002. *Parables for the virtual: movement, affect, sensation*. Durham, US: Duke University Press.

Morpurgo, M., 2010 [1982]. *War horse*. New York: Scholastic.

O'Reilly, B. and O'Reilly, C., 2004. *Hidalgo and other stories*. 2nd ed. Geneva: The Long Riders' Guild Press.

Randeria, S., 1999. Jenseits von Soziologie und soziokultureller Anthropologie: zur Verortung der nichtwestlichen Welt in einer zukünftigen Sozialtheorie. *Soziale Welt*, 50 (4), 373–82.

Reddy, W., 2001. *The navigation of feeling: a framework for the history of emotions*. New York: Cambridge University Press.

Shouse, E., 2005. Feeling, emotion, affect. *M/C Journal: A Journal of Media and Culture* [online], 8 (6). Available from: http://journal.media-culture.org.au/0512/03-shouse.php [Accessed 2 March 2015].

Therborn, G., 2003. Entangled modernities. *European Journal of Social Theory*, 6 (3), 293–305.

Tomkins, S., 1995. *Exploring affect: the selected writings of Silvan S. Tomkins*. E.V. Demos, ed. New York: Cambridge University Press.

The Turin horse, 2011. Directed by Béla Tarr and Ágnes Hranitzky. Hungary/France/ Germany/Switzerland/USA: TT Filmmûhely.

War horse, 2011. Directed by Steven Spielberg. GB/USA: Dreamworks SKG.

9   What's underfoot

Emplacing identity in practice among horse–human pairs

Anita Maurstad, Dona Lee Davis and Sarah Dean

Multispecies ethnography is a new field of study that rethinks the diverse and complex kinds of relationships that humans form with non-human animals (Haraway 2008; Kirksey and Helmreich 2010). Revitalized as topics of interest among anthropologists, focus in human–animal studies has shifted from viewing animals as objects to seeing them as subjects, as agentive individuals in engagements that matter – to animals, on the one hand, but also to human personhood and identities (Hamilton and Placas 2011; DeMello 2012). As Hayward (2010: 584), an advocate of multispecies ethnography, notes, animals should be viewed as acting on us in 'surprising and nuanced ways' rather than just as passive reflections of human intentions.

Writing extensively about the relationship with human and other animals, Haraway (2003: 16) contends that we are companion species, participants in on-going processes of 'becoming with' each other in naturalcultural practices. Speaking more directly to horse–human relationships, Despret (2004: 131) holds such entanglements as new articulations of 'with-ness', situations where species domesticate each other, and create new articulations of both speaking and being. Despret speaks of these engagements as anthropo-zoo-genetic practices, and engaging in anthropo-zoo-genetic practices, 'human-with-animal' and 'animal-with-human', respectively, are better categories to describe how beings alter or transform as a result of relating to each other.

A number of studies feature rider and horse as a pair and identify the roles of embodiment and bonding in developing a sense of partnership between horse and rider, a partnership that challenges hegemonic dualisms of horse as nature and human as culture (Wipper 2000; Game 2001; Höök 2010). Birke, Bryld, and Lykke (2004) also speak of horse–human practices as a co-creation of behaviour. Others grappling with a nature–culture divide feature styles of relating to or more grounded interactions with horses, such as a gendered mutual corporeality (Birke and Brandt 2009), embodied intersubjectivity (Brandt 2004; 2006), and the comparison of new American styles of natural horsemanship to British traditional horsey worlds (Latimer and Birke 2009).

Lesser focus is set on environment. Howes (2005: 7) notes that while the paradigm of embodiment implies an integration of mind and body, the paradigm of emplacement casts a wider net and 'suggests the sensuous interrelationship of body–mind–environment'. This makes sense concerning horse–human activities. Riding is experienced as vital and sensorial. Horses are certainly sensed, and catering to many senses, horse–human entanglements can be seen as 'sensescapes' (Howes 2005: 143), as sites where experiences are produced by particular modes of distinguishing, valuing, and combining the senses. Senses are experienced with body and mind, but they are also literally placed – *on* horses, *with* them, and *in* different contexts, be it landscapes, stables or other. Bringing the environment into the analysis, allows interweaving the sensorial with the material, says Howes (2005: 7): it allows us to 'reposition ourselves in relationship to the sensuous materiality of the world'. Moreover, bringing materiality into the equation allows seeing horses and humans as relational materialities, as bodies and minds attuning to each other through various processes and performances. Horse and human 'become' together through engaging in activities that affect both parties. Relational materialities is a concept brought to the fore by scholars in the sociology of science. Particularly, actor network theorist John Law (1999) holds that materiality is an effect, an outcome, of work performed in networks made up of human and non-human elements. And it is the relations between the entities that decide form and attributes of material elements. These classical actor network theory perspectives are not commonly used in human–animal studies, nor in multispecies ethnography studies, but Thompson (2011) is an exception. Focusing on two equestrian activities related to the Spanish bullfight, and how they are emplaced and mediated through various technologies, she points to crucial differences concerning horse and human in the two situations. Their abilities to perform as pairs and agentive individuals differ, and Thompson's (2011: 248) point is that material arrangements, technology, mean something to 'centaurability', i.e., the embodied feeling of being one and acting as one: 'the centaur emerges from particular practices, processes and performances'.

Our multispecies ethnography study on human–horse engagements will weave the concept of emplacement together with a focus on somatic attuned practices, and domestication as emplaced sensuous and relational materialities. In what follows, human and horse engagements will be analysed as experientially situated in three different kinds of emplacements: First, we address situations of being *on* the horse. Rider and horse bodies are placed, literally flesh to flesh. They exert an emplaced and embodied communication across a species divide, a communication that has effects on bodies and minds (Maurstad *et al.* 2013). Second, we turn to situations where riders speak of being *with*. Riders' narratives about horse–human relationships and bonding include many situations of just being together with the horse. When not riding, different mind and bodily work come into play. And environments matter. Riders care about the contexts in which their engagements with horses are emplaced, and they are being affected by elements therein. Third, we address being *in* different terrains and sports. Focusing on the equestrian sports of Icelandic gaited horse riding, dressage, and endurance riding, we demonstrate how emplacement in different terrains engender a sense of mutuality between horse and rider, not as subject and object but as two intra-active, agentive individuals (Davis *et al.* 2013). Finally, we discuss how horse and human are paired together, defined and distinguished, through complex modes of emplaced attention and attachment, involving somatic modes of attention (Csordas 1994; 2002), as well as cognition and affect (Despret 2004). Horses are soul mates, but also body mates to many humans, and the relationship is one that affects and defines both parties. Through emplaced practices, the species domesticate each other. They are trained together, as materialities emplaced in intra- and interactions with effects on the species.

The study is based on narrative data collected in over 60 interviews with riders in North Norway and Midwest United States.[1] We used an open-ended interview format, grounding riders' own reflections of how they relationally portray and enact themselves and horses. Riders participate in a range of equestrian activities: Gaited horse riding, endurance, eventing, dressage riding, and hunter-jumping.[2] The authors of this article are, similarly, as different in age, geography, and sport, as our informants. We also keep horses and have done so for a number of years, affording us an insider knowledge and familiarity with the field (Brandt 2004).

Multispecies ethnography is, to an even larger extent than other ethnographic efforts, faced with the problem of representation. Although we share their company on a daily basis, no horses were interviewed in our study; it is their humans that speak on their behalf. The common-sense experiential worlds revealed in informants' narratives, however, in what Quinn (2005: 2) calls 'culture in talk', show how dualisms of nature–culture, control–mutuality, and object–subject are transgressed and rejected as radical separations in kind. When speaking about observations, experiences, and daily practices, riders reveal a variety of practices with horses that offer new insights into how the species domesticate each other through emplaced activities.

The local environments and sport communities that riders belong to are small. In order to not jeopardize anonymity, we have chosen to leave out contextual information about who the interviewees are. Riders chose their own fictional names, which are utilized within quotes.

Emplaced on horses

Riding is a practice where horse and human bodies communicate through a set of cues and signs. Emplaced on the horse, body weight and position on the horse, as well as slight pressures from reins and legs, are basic tools to signal a rider's wish. Iterative training is needed for the human to learn how to signal correctly to the horse and for the horse to understand, but when both horse and human are well trained they attend to the feel of each other and respond appropriately. The signals are based on ethological ideas about what is natural to the horse, but they need to be learned by horse and rider; they are new material-semiotic practices (Birke *et al.* 2004).

Rider Urdur stresses that it is a real challenge to learn new ways of body behaviour: 'it is a real demand on a sort of self-upbringing to succeed'. The new body behaviour is mindfully thought about, before being applied. Emerita speaks to some of the things a rider is bearing in mind – always:

> Riding is, if I am to sit and ride a horse I want to do it in the right way so that I do not destroy the horse. [...] You are to train it so that it can last a whole life with me on his back. [...] This is why we are sitting straight, head resting on the body; we are not leaning to the side and doing those silly things, we try to make it easy for the horse to carry us – so that we do not destroy it.

In addition to these physical aspects, Emerita says that training involves mental work. The horse needs to be relaxed, to find his balance with the rider before exercises are rehearsed, she says. Training must be performed with attention and care involving body and mind:

> When I decide to do a thing, I first think it, then I start to do something with my body, and then I must let the horse have a chance to get it into his brain. [I must] not rush, be patient, give him a chance to get it, to solve that task.

Emerita reflects on materialities that meet and attune. Thomas Csordas (2002: 7–8), a central scholar establishing perspectives on embodiment as analytical focus and paradigm, speaks of 'somatic modes of attention', defined as 'culturally elaborated ways of attending to and with one's body in surroundings that include the embodied presence of others'. Csordas (1994: 4) refers to bodies as lived practices, as actors in the world; bodies have histories and are as much cultural phenomena as they are biological. These perspectives are relevant for analysing engagements that affect both horse and human bodies. Riding is a situation where bodies are closely involved, and horses have sensitive bodies. As philosopher and horse trainer Vicki Hearne (2000 [1986]: 108) points out: 'Every muscle twitch of the rider will be like a loud symphony to the horse'. Humans learn to play softer symphonies as they develop skills. They are balancing according to a feel of the other, the horse, attuning their bodies to sensations of the horse bodies. The horse too must attune to the human body, developing the necessary strength and balance in order to carry a rider and perform his or her wishes. In essence, 'on horse' is a place being transformed through the meeting between 'sensuous materialities' (Howes 2005: 7). It is a sensed place, sensed and transformed.

There are many senses involved. Analysing riding through perspectives on rhythm, particularly the sport of dressage, Evans and Franklin (2010: 184) hold that riding is 'synchronization of eurhythmic bodies of horse and rider'. Riding is embodied learning and 'production of equestrian bodies, both human and non-human' (Evans and Franklin 2010: 185). They hold that this training of bodies produces trust: 'this partnership is formed through repetition, through the rhythmic application of embodied incremental training, the ultimate purpose of which is to produce what in the equine world is called 'partnership' and 'trust'' (Evans and Franklin 2010: 184).

Our informants do speak of rhythm. Halla says that: 'Part of the whole riding is the rhythm of it; the feel of the rhythm of it whether it's the trot or the canter. The horse breathes rhythmically and it's all tied together'. It should be noticed here that gaits are rhythmic events: There is a clear ga-de-dung, ga-de-dung, ga-de-dung to the three-beat canter, and a clear black-and-deck-er to the four-beat tölt.[3] When learning these gaits, embodying the rhythm is a crucial emplaced activity. In dressage, for example, knowing (feeling) where the inside hind foot is, and what it is doing, is essential to mastering even the most elemental movements of the sport. Discussing places as sensed, Feld (2005: 185) maintains that 'understanding the interplay of sound and felt balance in the sense and sensuality of emplacement, of making place' is important. While Feld discusses sensing territories, his comment is very valid for horse–human activities.

An interesting point concerning how tangible and intangible materialities are interwoven is brought up by rider Rebecca. In situations of being *dis*placed from horses, Rebecca speaks of rhythm as something she longs for: 'I miss the rhythm of moving together, I miss the outside benefits of health and I miss that relationship of having an animal that just is gonna love you back, just unconditionally'. Rebecca is connecting rhythm and love, mixing motion and emotion, and the two are crucial elements of everyday activities, of moving in landscapes, says Feld (2005: 181): 'Because motion can draw upon the kinaesthetic interplay of tactile, sonic, and visual senses, emplacement always implicates the intertwined nature of sensual bodily presence and perceptual engagement'. Love of horses comes from (among other things) the rhythms the pairs produce. There is a materiality to place and activity as heard and felt, consisting of both tangible and intangible elements.

A further way of elaborating how motion and emotion can be seen as linked, producing sensorial materialities, is found by looking to a goal that Emerita and many others speak to – moments of being in sync: 'it is just so good when you feel you are one with that horse and it just floats on and you melt together and it is just perfect. Because that is what you feel in golden moments, that, gee, now it is as good as it can ever be'. In fact, Rebecca says that her very reasons for riding are such in sync experiences, and she elaborates: 'the feeling when you and your horse are in sync and everything that is communicated is fluid and it just, everything works out like, like you're one, you know'. Urdur also points to being emplaced in the environment. She enjoys: 'the feelings of becoming one with the large animal in nature'.

Such moments are characterized, as the riders say in these quotes, by a thrilling feeling of ease, of being one, and the centaur myth, part human part horse, is often used as a metaphor for such events (Game 2001). We have already pointed to Thompson's (2011) study, which focuses on how 'centaurability' is mediated by technology. Other scholars also discuss these in sync experiences that riders talk about. Argent (2012: 120) suggests, as do also Evans and Franklin (2010), that the thrilling feelings of in sync experiences are explained by the performance of rhythmic synchronic movements, an interweaving of motion and emotion. Emplaced on horse, humans can feel the sensorial experience of being one. The thrilling feelings are so strong that 'It is no longer the doing, but the being, that becomes important' (Argent 2012: 121). And Argent holds that horses are in for it too. They are social creatures with characteristics similar to humans in that they form cooperative alliances. Thus, she suggests that it is not only humans that value synchronized corporeal behaviour, but that horses, too, take pleasure from these emplaced intra-activities. Also Evans and Franklin (2010) suggest that horses enjoy harmonic motion, with reference to that they too are creatures with senses of rhythm.

Emplaced with horses

Speaking of being emplaced *with* horses, that is, sharing the same ground underfoot, rider Ola can set the scene:

> As I see it, riding is just part of having a horse... if you are to be pleased with your riding, and the horse when you ride, I think you need a very good relation before you ride. You must be ready to ride, mentally. You cannot just jump on a horse and ride off and think that you, that this will work well. I think the riding starts at completely different places.

Ola is specific about places – there are different places where one interacts with horses. Riders do a lot of work *with* horses, in places where various less-focused sensations are present. Body awareness on how to perform when both human and horse have the same footing, is still present. Bodies need to learn how to perform with horses emplaced in more settings than on the horse, as Becca notes, reflecting on body awareness in various places: 'It's not just your being in the saddle, it's how you are with them at feeding time, it's how you walk them through the gate'.

Becca's statements speak directly to emplaced bodily actions – with horses. Riders are emplaced in many different situations while being with the horse. They are in the stall, at the field, in the paddock, but all the time they are being with, and learning how to behave with, and communicating with this other species in situations other than sitting on them. Being emplaced in these 'being with' situations, riders must attend to feel, in ways that involve body and body control. Rider Aurora speaks of a shy horse she once owned and how she had to relate to this particular individual in certain corporeal manners in order to have a dialogue with him. She talks of herself as somewhat whimsical by nature, but this is something she had to control when being around the horse. She changed her body language in one important aspect: 'I did not wave my arms'.

This is perhaps not a major bodily change, but it is important that it is the horse natureculture that sets off these responses in Aurora's body and mind. The horse is domesticating Aurora, re-training her physical attributes so that she becomes a different person, one with a more still personal behaviour. She adapts to new naturalcultural practices from meeting with this horse.

Emplaced in other situations, grounded with the horse, there are more senses that come into play. Urdur enjoys sitting with the horses when they eat. She enjoys watching and listening to them and says it is soothing. She also contrasts it to watching a dog eat, which is not pleasant to the same extent. Bernadette speaks eloquently about the aesthetics of the horse – she loves to watch their faces, ears, nostrils, and several riders say they love the smell of horses. Similarly, Deany notes: 'I was breathing it in and I could almost start crying [...] With riders, just being around the horse, smelling what a horse smells like, that just captivates you'.

Grounded with the same footing as horses is also something that riders think affects the horses. It suits riders, but also horses. Katla is sleeping in the stable to become better acquainted with a new horse that she is going to train – ensuring trust to the horse being stabled by being with her. There are also elements of care and feedback from care. Henriette speaks of being happy that her horses are well taken care of: 'I feel that everything about them, being well taken care of, having a nice time, seeing to it that they get their food, that we take out the pooh, that they are in an environment that is good, all that is taken care of'. Henriette keeps her horses where she lives and place becomes particularly valued to her: 'Every time I go out, they are there, coming from work they are there, it is recreation beyond everything'. She likes being with them, in their corral. She enjoys brushing them, and they are clearly happy that she is there, she says. To Henriette then, the paddock is a site where she and horse are emplaced together, doing body–mind–environment interrelated work of importance for their relationship and, we could add, their materialities – who they become, as horse and human.

We can let Deany sum up how horses, materialities, emplacements, and attuning to are interwoven. Dealing with horses, she says: 'became a happy place and a happy thing for me to do'. Indeed, the narratives of riders reveal that the horse–human relationship holds very sensuous materialities interacting. Being with is of so much importance that many riders make plans for elderly life, where they might be displaced from horse riding. This is the case with Bonnie, who wants to be wheeled out to the stable if she gets old and disabled. She has told her daughter:

> listen, if I get in a wheelchair or have Alzheimer's, bring me out to the barn. I might not even remember my horse. Really, I'm serious. If I can't get around, if I can't clean stalls, bring me out there. Wheel me into the barn.

With Despret (2004) we can say that the engagements that horse and human perform create a 'human-with-horse' that is different from a human without one. There are generic aspects that matter; engaging with a horse is different from being with a dog. The with-ness with a particular horse leads to a co-creation of new articulations of being human. And places for engagements matter. Humans engage their bodies and minds with horses in a range of different environments, both factual and not factual, an example of which being an imagined future Alzheimer diagnosis.

Emplaced in terrain

As we conceptualized this study of horse–human relationships, we did not anticipate or recognize the thematic salience or importance of terrain in the narratives until we began to pool our data. Riders are *on* their horses riding, and *with* them when feeding and watching, as we have discussed. But they are also *in* different terrains, and terrains covered, as well as shared equestrian sport cultures, constitute emplacements of horses and riders that demand, select for, or elicit similar skill sets and responses from horse and human.[4]

Lynn and Missy tell us how dressage horses and their riders become mutually and reciprocally attuned, focused, and in-touch as they micromanage their way moment by moment over well engineered, predictable, and geometrically detailed terrains, clearly demarcated by visible letters surrounding the arena, and invisible letters within it. The terrain for dressage is an arena, 20 by 40 or 60 meters, where various dressage exercises are performed with subtleness.[5] It is a rather contained physical habitat, but to say that the terrain or ground is unimportant misses the fact that it is supposed to be well groomed to exacting standards, in order for the horse and rider to focus attention on the task at hand. Within a predictable and clearly marked arena, horse and rider as deep thinkers can concentrate on each other and not be distracted by whatever goes on in the immediate environment outside of the arena.

Dressage riders speak of themselves and their horses as very focused, as, for example, Lynn notes:

> I like the intensity – that focus factor – that thinking where I have to put her haunches, what foot is striking and is she bearing her weight correctly? Am I shifting my weight correctly? I like that feeling of being focused on what I'm doing – as opposed to eventing where you have to trust the horse to do it.

Missy is also talking of intensity. She tells about a time she was entering the ring, riding toward the judges and they were waving their arms, the audience was shouting 'stop!'; a storm had started and neither she nor the horse had noticed it.

With the dressage arena as footing, mutual focus, intensity, attunement, interiority, and micromanagement emerge as words that dominate dressage riders' narratives about the horse–human relationship. There is a kind of bounded interiority to the sport. Horse and rider, interacting and moving together, are the environment. The physical world outside the pair demands attention, because where you are and where you are going in the arena counts.

Endurance riders are different. They are rugged individuals (Cowles and Davis 2013), like Hot Shoe Sue and Sky, who relate how their horses log mile after mile over varied and challenging terrain, where energy, stamina, conditioning, and stoic endurance guarantee survival, if not winning, for both human and horse.

Being in partnerships with horses is an essential part of the endurance experience, but so is being in the environment. Endurance riders train and condition their horses to navigate challenging and sometimes dangerous terrain. Endurance riders and their horses are expected to be strong and stoical to endure and log many miles together. Hot Shoe Sue elaborates:

> Endurance riders […] it's not a sport for whiners. There is congeniality on the trail. The horse is trained as an athlete that does lots of miles. It's not a public sport. It's people who want to see the country and you don't need to dress up or mind being wet or tired. It's finishing, not winning. It's the guts and glory of riding over terrain that is not in the best of conditions. It's a survival sport, not a beauty pageant. It's not so much an issue of skill as it is you and the

horse versus the trail and it all comes down to how you manage your horse and what kind of animal he is.

Sky also says that endurance is natural to the horse: 'I like endurance because you're out in a natural area and you're doing what comes natural to the horse – not forcing them to do something that's not natural like reining or jumping. This is something they like to do'. Hot Shoe Sue agrees:

> If a horse likes to go forward they will do this. Most of the horses that do this are successful and enjoying what they do, and they'd rather do it than stand around in a stall and go into an arena for an hour or two a day.

Miles logged, rugged individualism, heroic stoicism, silent suffering, hard work, and the ability to function in a risky environment are portrayals given to both human and horse in this sport. These are values they hold to be quintessentially American. Yet many of these views characterize the North Norwegians and their narratives of the Icelandic horse.

The hardy Icelandic gaited horses and their all-weather, hardy Norwegian riders, like Urdur, Kane, and Ola, bring training into a mixed and versatile outdoors, where together they relax, breathe in the fresh air, and open up their selves and senses and become immersed in the region's distinctive nature. Urdur explains this in some detail:

> Yesterday I experienced a wonderful trip. It was warm. It was sultry damp, the forest was damp from nearly tropical scents. I took a long tour around the island and explored a new area in the north. The seashore… you can have all the nuances. We cantered a bit and tölted a bit, and we just walked through the forest. Of course, you could experience this on foot, but you could not reach such great distances as with a horse, plus one would not have the feeling that one is together with the animal on that journey.

To Urdur it is of great value to be able to see the landscape through the eyes of a rider. She had thought about this, long before she got her own horse. She enjoys being alone with the horse in nature, meeting other animals and birds in the forest. And as in the case of Kane, she notices that the horse plays a major role in this. Kane speaks to both embodiment and emplacement when being in the forest and jumping over a creek. He enjoys riding on his own in the forest, a moment when he is: 'happy with the horse, can feel him working, having contact, I feel he is trusting me, we are to jump over a creek, or going down a steep hill, then he trusts me'. Kane also says that the experience of nature is stronger when experienced through the horse. Talking about riding on tours he says:

> You look for the experience, and that is much stronger, through the horse, you experience through the horse, experience the nature in a very special way, and that does something with me, that is for sure […] that feeling is very

strong, I do not need to have it very often, but I must have it now and then. It is extremely wonderful.

Like the two other equestrian groups, gaited horse riders also include the animal in the descriptions of their practices. The Icelandic horse likes to run, to stretch, and to be in nature. The horse enjoys the variation, having different types of ground underfoot. As Ola says:

> I seldom ride the same routes. It's much more positive to ride different places. It's very fun to herd sheep or cows in the forest. You never know where you might go. Horses like that. They like the physical challenges to work themselves a bit.

In the Norwegian narratives nature becomes more elaborately themed – not so much as something to be mastered, but something to be a part of, to be experienced and felt. Among the Norwegians the outside is brought into the inside and vice versa.

Emplaced in domesticating practices

Thinking analytically through the concepts of materiality and emplacement brings valuable nuances to the analysis of horse–human entanglements. We have demonstrated complex ways in which emplaced on, with, and in horsey situations affects a series of interwoven constructions of shared ecologies of horse–rider relations, identities, personhoods, and psyches. Comparative analysis of discourse from riders with different ground underfoot reveals thematic differences that distinguish the practitioners. Horse and human are paired together, defined, distinguished, and identified by their activities. Materialities are produced in sensuous ways; it is through sensing each other, through motion and emotion, that horse and human develop.

Co-being in engagements, be it on, with, or in various environmental entanglements with horses, both horse and human learn new definitions of being. It should be stressed that, although their communication is built upon cues that the species understand, it is also about establishing something entirely new. It is a genuine naturalcultural exchange, a domesticating practice where horses' natureculture meets with humans' natureculture, and the negotiated outcome, the new naturalcultural practices, work within frames of mutuality and trust.

Evans and Franklin (2010) note that riders and horses make equine landscapes. We have shown that different landscapes make different horses and riders. Moreover, the concept of emplacement brings focus to how place is more than landscape. There are places for feeding, for grooming, for just seeing and listening – a range of places that matter to the development of relationships between horses and riders. And as relationships are about attuning to each other, the flip side of this is that there exist a range of places where horse and human are being domesticated. Domestication is a useful concept here because it underlines that human and animal change as a result from their meeting. Not only do animals adapt to human intentions and wishes – our study shows how humans adapt in order to communicate and have a relation to horses. Being on, with, and in horsey situations has effects on bodies. Riders' bodies perform differently in society than non-riders' bodies do. Communicating with horses, riders' body kinetics change to create bodies that are mindfully controlled in their talking. In addition to the obvious physical aspects, like growing new muscles in legs, butts and other parts of the body, riders learn to understand that first, they have talking bodies, and second, how to talk to horses through them. Through somatic attuning (Csordas 2002), communication develops in ways where action and response lead to understanding the other in more nuanced ways.

There are also effects on personalities: Niki has developed a better understanding of 'being able to understand people even if it's not verbal'. And Barbara states that horse experience translates into the kind of person she is in several ways. She has to be centred inside, confident, assertive, fair, and 'all that translates into human relationships'. She has become 'more open to feeling, emotions, relationships now'. It is a real transition that riders learn more about being human, through being with horses. Ajay sums it up: 'because you see things through a horse's eyes I think it just changes your view on the whole world'.

These changes that the riders talk about are linking relational materialities, emplacement, and domestication. The horse as sensed material place invites motion, emotion, and attuning to – on horse, with horse, and in situations with different ground underfoot. There is stillness, there is moving to and from, and within, the places. Riding is moving from place to place, but also being in one place all the time, on the horse. Place, on a horse, is not static, but it is fluid: muscles and senses are engaged and transformed in ways that riders elaborate upon. Riding (or as Argent [2012: 124] states 'riding with') becomes embodied interaction that includes interrelationships with the environment (see Howes 2005). This provides a more nuanced understanding of what constitutes the 'it' in horse–human relations. Horse–human relationships are emplaced activities, emplaced body–mind work, in environments, with sensuous and material effects.

Laura's sensations of place bring a nice conclusion to this chapter. Speaking of moments where intrinsic sensations play out, she focuses on mornings at the stable, when 'the sun is hitting just right through the windows' emplacing her in a special emotional state, as such 'beautiful golden sunrises […] makes everything seem like a dream'. To Laura this horse–human entanglement provides a sensuous start of the day. Horse–human entanglements hold many such sensuous situations, sensuous materialities, tangible and intangible, that have effects on both species.

Notes

1 Anticipating that our data would bubble from the ground up and reflect culture in talk (Quinn 2005), we all asked the same general questions. They were: 1) Why do you ride?; 2) Tell me about your life as a rider.; 3) How does this relate to the kind of person you are?; 4) How does riding relate to other aspects of your life?; and 5) How is your experience the same or different from the experiences of other people? There were 52 women and 8 men, aged 20–70, in our sample. Interviewees were recruited at a variety

of venues. In addition to competition events and clinics, we interviewed riders in their homes, at local barns, and at riding facilities. We also recruited participants among our common-interest friendship groups. Interviewees were approached by personal contact, by phone, or face-to-face. All interviews were recorded and later transcribed. Transcriptions of interviews ranged from 10 to 28 double-spaced pages. Anita translated interviews she conducted from Norwegian to English. We are all accomplished riders; Dona in dressage and eventing, Sarah in endurance riding, and Anita with Icelandic gaited horses. The period of data collection extended from summer 2011 to spring 2012.

2 Dressage riders perform a series of prescribed movements, including gait transitions in a fenced arena. Highly valued is the appearance of an effort-free rider with a horse willingly performing the requested movements. In eventing horse and rider perform with stamina and toughness in three disciplines: dressage, stadium-jumping, and cross country, the latter being the segment that defines the sport. The cross-country course consists of solidly built fences. Endurance riding is long-distance trekking over diverse terrains. Enduring distance, terrain, and weather is what defines both horse and rider. Gaited riding is a particular sport for Icelandic horses and their riders. In addition to the walk, trot, and canter, the Icelandic horse also does tölt and pace. Rhythm, cadence, speed, and action are what define the praised horse and rider pairs. Hunter-jumpers jump fences in an enclosed arena.

3 Icelandic horses are gaited horses. Four-gaiters do *tölt* in addition to the walk, trot, and canter. Five-gaiters do an additional gait called *pace*. Tölt is a four beat gait, like the walk, only much faster. Pace is a very speedy two beat gait.

4 This theme is more fully elaborated in Davis *et al.* (2013).

5 Evans and Franklin (2010) describe the dressage sport in more detail.

References

Argent, G., 2012. Toward a privileging of the nonverbal: communication, corporal synchrony, and transcendence in humans and horses. *In:* J.A. Smith and R.A. Mitchell, eds. *Experiencing animal minds: an anthology of animal–human encounters.* New York: Columbia University Press, 111–28.

Birke, L., Bryld, M., and Lykke, N., 2004. Animal performances: an exploration of intersections between feminist science studies and studies of human/animal relationships. *Feminist Theory,* 5, 167–83.

Birke, L. and Brandt, K., 2009. Mutual corporeality: gender and horse/human relationships. *Women's Studies International Forum,* 32, 189–97.

Brandt, K., 2004. A language of their own: an interactionist approach to human–horse communication. *Society and Animals,* 12 (4), 299–316.

Brandt, K., 2006. Intelligent bodies: embodied subjectivity in human–horse communication. *In:* D. Waskul and P. Vannini, eds. *Body/embodiment: symbolic interaction and the sociology of the body.* Aldershot, UK: Ashgate 141–52.

Cowles, S. and Davis, D., 2013. Krusty and other sexagenarians: heroic self-stylings of aging among equestrienne time rebels. *Teaching Anthropology: SACC Notes,* 19 (1 & 2), 29–34.

Csordas, T.J., 1994. Introduction: the body as representation and being-in-the-world. *In:* T.J. Csordas, ed. *Embodiment and experience: the existential ground of culture and self.* Cambridge, UK: Cambridge University Press, 1–24.

Csordas, T.J., 2002. *Body/meaning/healing.* New York: Palgrave Macmillan.

Davis, D., Maurstad, A., and Cowles, S., 2013. 'Riding up forested mountain sides, in wide open spaces, and with walls': developing an ecology of horse–human relationships. *Humanimalia: A Journal of Human/Animal Interface Studies* [online], 4 (2), 54–83. Available from: www. depauw.edu/humanimalia/issue%2008/davis%20et%20al.html [Accessed 5 May 2015].

DeMello, M., 2012. *Animals and society: an introduction to human–animal studies*. New York: Columbia University Press.

Despret, V., 2004. The body we care for: figures of anthropo-zoo-genesis. *Body and Society,* 10 (2–3), 111–34.

Evans, R. and Franklin, A., 2010. Equine beats: unique rhythms (and floating harmony) of horses and riders. *In:* T. Edensor, ed. *Geographies of rhythm: nature, place, mobilities and bodies*. Farnham, UK: Ashgate, 173–85.

Feld, S., 2005. Places sensed, sensed places: toward a sensuous epistemology of environments. *In:* D. Howes, ed. *Empire of the senses: the sensual culture reader*. Oxford, UK: Berg., 179–91.

Game, A., 2001. Riding: embodying the centaur. *Body and Society,* 7 (4), 1–12.

Hamilton, J. and Placas, A., 2011. Anthropology becoming...? The sociocultural year in review. *American Anthropologist,* 113 (2), 246–61.

Haraway, D., 2003. *The companion species manifesto: dogs, people, and significant otherness*. Chicago, US: Prickly Paradigm.

Haraway, D., 2008. *When species meet*. Minneapolis, US: University of Minnesota Press.

Hayward, E., 2010. Fingereyes: impressions of cup corals. *Cultural Anthropology,* 25 (4), 577–99.

Hearne, V., 2000 [1986]. *Adam's task: calling animals by name*. New York: Akadine Press.

Höök, K., 2010. Transferring qualities from horseback riding to design. *In:* A. Blandford, J. Gulliksen, E.T. Hvannberg, M.K. Larusdottir, E.L.-C. Law, and H.H. Vilhjalmsson, eds. *Proceedings of the sixth Nordic conference on human–computer interaction,* 16–20 October 2010 Reykjavik. New York: ACM, 226–35.

Howes, D., ed., 2005. *Empire of the senses: the sensual culture reader*. Oxford, UK: Berg.

Kirksey, E. and Helmreich, S., 2010. The emergence of multispecies ethnography. *Cultural Anthropology,* 25 (4), 545–76.

Latimer, J. and Birke, L., 2009. Natural relations: horses, knowledge, technology. *The Sociological Review,* 57 (1), 1–27.

Law, J., 1999. After ANT: complexity, naming and topology. *In:* J. Law and J. Hassard, eds. *Actor network theory and after*. Oxford, UK: Blackwell, 1–14.

Maurstad, A., Davis, D., and Cowles, S., 2013. Co-being and intra-action in horse–human relationships: a multispecies ethnography of be(com)ing human and be(com)ing horse. *Social Anthropology,* 21 (3), 322–35.

Quinn, N., 2005. Introduction. *In:* N. Quinn, ed. *Finding culture in talk: a collection of methods*. New York: Palgrave, 1–34.

Thompson, K., 2011. Theorizing rider–horse relations: an ethnographic illustration of the centaur metaphor in the Spanish bullfight. *In:* N. Taylor and T. Signal, eds. *Theorizing animals: re-thinking humanimal relations*. Leiden, Netherlands: Brill, 221–53.

Wipper, A., 2000. The partnership: the horse–rider relationship in eventing. *Symbolic Interaction,* 23 (1), 47–70.

PART III

From objects to subjects

Exploring animal subjectivity

10 Moving (with)in affect

Horses, people, and tolerance

Lynda Birke and Jo Hockenhull

At this my hert is astonnied, and moued out of his place (Coverdale 1535, qtd in *Oxford English Dictionary* 2015)

It's such an emotive journey and on that journey you take along your best friend and they experience it too (interviewee in Birke 2008: 116)

To move: this verb has myriad meanings. We move through life, we relocate, we move together or apart, things move us – either as physical transport through space, or emotionally. Moving as relocation and as affect are not only linked lexically; Coverdale's use of 'moued' [moved] evokes emotions as well as an imagined displacement of his astonished heart. Similarly, the horsewoman quoted above puts stress on moving through life and affect, on an 'emotive journey' of discovery with her horse.

Feeling an emotional connection is, for most of us, a crucial part of living with companion animals; animals do, indeed, sometimes seem – at least to us – to become our best friends. Even though many animals suffer terribly at human hands, there is the possibility of more positive interactions with us, for at least some animals (such as companion animals), and perhaps even the possibility of love (see discussion in Cudworth 2011). But in whatever ways they experience the relationship, there is no doubt that they can move (and perhaps be moved by) us – emotionally, and physically.

In this chapter, we draw on various studies of human–animal relationships – in particular, from our studies of humans and horses. We focus on how horse people understand, talk about, and explain their horses, within the context of specific affective relationships. Building a working relationship between horse and human involves learning to work together, to have expectations about how the other moves and behaves (Argent 2012; Birke and Hockenhull 2015). But this relationship – when it works reasonably well – also configures and permits (some) less predictable actions within it, a kind of breaking out of acceptable bounds on the part of the horse. Sometimes, horse owners[1] find such behavioural challenges unacceptable, and will struggle to 'make' the horse behave as they wish. Others are willing to tolerate the horse's challenges to an extent, as we will argue. What, then, are the contexts in which we tolerate these challenges – or not? And how do owners interpret their horses' reactions? What do these stories say about human–horse affective relationships?

Human–animal relationships: understanding/explaining the other

Relationships between us and companion animals necessarily involve learning to understand the behaviour of the other, if only partially. There may be dire consequences for people or animals from a failure to read the other's intentions. Getting it wrong can have very serious consequences for humans (fatal bites, for instance) – but it very often has tragic consequences for the animal.[2] If animals misinterpret human intentions (or humans are just plain cruel), and react accordingly, then they are likely to be mistreated, or killed.

In the eyes of non-human animals, we humans may be many things. We might be of no immediate interest, we may be prey or predator, depending on the species; we might at times seem bizarre or stupid, or a source of curiosity, or of food or shelter. We may, for many, be a source of danger, of potential threat. Or we may, sometimes, be seen as a friend or collaborator, someone with whom an affective bond can be built. For companion animals such as dogs or horses, our presence can sometimes (seem to) be welcomed (here comes my dinner!), or not (groan; the bridle means work, that crate means a trip to the vet). To establish some kind of relationship, both human and animal must be able to read the other's intentions, at least to some extent. There is plenty of evidence that some companion species, having spent millennia in close proximity to us, have evolved understanding of human gestures – the meaning of pointing, for example, or the direction of our gaze (e.g., Schwab and Huber 2006; Proops and McComb 2010; Topál and Gácsi 2012). Getting it right, more or less, can make the difference between an affectionate head-rub or a kick.

How, in turn, do we understand companion animals? Partly, that will depend on experience – with animals in general, with that particular species, with that individual: out of such experience we can build expectations of how that animal might react. But humans also draw on wider cultural beliefs about different species, as well as ideas about how animals 'should' behave (e.g., during training). One widespread (and highly contested) idea in popular culture is that we should assert domination, to tell the animal 'who is boss'. To some extent, of course, domination is intrinsic to the very act of keeping/living with companion animals, as they are always within our power (Tuan 1984; Cudworth 2011). Such structural domination does not necessarily mean that individual relationships cannot flourish; indeed, Cudworth (2011) points out that, despite the clear differences in power, affectionate relationships between companion animals and their people are possible.

However, disparities of power go hand in hand with using techniques of domination in attempts to control animals. According to some animal trainers, people need to think of the animal as 'wild' (the 'wolf within' the family dog, for instance: van Kerkhove 2004), so that the human becomes like a 'pack leader', who must put the animal 'in place'. Owners thus speak of the animal 'needing to respect' the person. Writing about interactions with her dogs, Barbara Smuts (2006) considers the advice given by various dog trainers, who suggested that the dog 'needed' to be dominated – a theme very familiar to anyone who has worked with horses (see also Argent 2012). Luckily for the dogs, she took a different view, and spent time watching how they interacted with other dogs as a way of trying to figure out how they related socially, so that she could work with them to develop ways of relating. A crucial lesson to be learned from this, she argues, is that these dogs' stories

> challenge our tendency to think of individuals as primary, and relationships as secondary phenomena that are 'caused' by the actions of individuals [...] [for] highly social species like humans and dogs, it is impossible to understand an individual's behavior outside the nexus of their most important relationships (Smuts 2006: 123–4).

For companion animals, those important relationships often include humans. Alongside the figure of the 'wild' animal, in 'need' of domination, who must learn to 'respect', there is another cultural trope – that of the animal as potential friend, part of that nexus of relationships. Smuts's dogs are clearly much more than domesticated animals 'needing' to be tamed: they are friends, with whom she establishes unique relationships. We are willing to tolerate the occasional waywardness or idiosyncrasies of friends, whoever they are (as many dog people will testify!), and will acknowledge their active contribution to the relationship. It is this idea of the animal as friend to which the quotation above refers: taking along your 'best friend' on a journey of discovery about each other was, for that horsewoman, a profound experience. Friends, moreover, sometimes offer us help – and many animals actively do. The idea of animals as 'helpers' is widespread in our culture, occurring in many contexts, from representations of lab animals, portrayed in advertisements as 'helping' medical progress (Arluke 1994), to 'helpful' animals in cartoons and folk stories, to stories of animals who go out of their way to assist their human (see McElroy 1996). Horseriders, too, often refer to their animals as 'helpful' (Thompson and Birke 2014). This is not to say that the animals actively think about offering help (though they may do!); rather, it is to say that, in close relationships with humans, their ability to assist, to work with the person, becomes integral to the relationship.

Culturally, we draw on many understandings of companion animals in our interactions with them. The twin tropes of 'wild animal within', who might be held in check, and 'helper/friend', who can work alongside, undoubtedly configure our affective relations with companion animals. At times, in our dealings with other animals, we may draw on one trope more than the other – although we also often wish to see companion animals as friends while simultaneously wanting respect from them.

The rest of the chapter focuses specifically on horses and how our understandings of these animals' behaviour play out in affective relations with them. While ideas of the 'wild within'/domination are evident in horsey worlds, the idea that such animals can be friends prevails alongside, particularly among people who keep horses for leisure pursuits (Birke 2007; Birke *et al.* 2010). And key to seeing animals as friends is attempting to understand where they are coming from, why they do what they do.

Significant others: horses

Horses have many different meanings within our culture, and carry enormous symbolic significance. Historically, they have been agricultural workers, means of transport, and unwitting conveyances of war. As those uses declined, horses became more often used for leisure pursuits (Leckie 2001). But, they are increasingly seen as friends or companions, with whom a close bond may be forged – and sometimes they become 'significant others' with whom people are willing to spend considerable time.[3]

Within equestrian worlds, however, horses have traditionally been seen in rather utilitarian ways: they are supposed to 'obey', to do what is required by the human (cf. the domination model noted above). Training aims to ensure that the animal learns 'obedience', moves in specific ways, and performs required tasks. How individual horse owners understand this depends, of course, on circumstances and experience. Some may respond with punishment, others with encouragement, if the horse does not 'get it right' (Hockenhull and Creighton 2013). Associated with the shift toward seeing horses as friends, however, there is growing responsibility to abandon or minimize ways of interacting centred on 'discipline', and instead encourage ways of achieving cooperation through kindness and building relationships (though this perception can still include expectations of 'respect' from the horse: see Birke 2007).

Despite that shift, 'disobedience', or wayward behaviour, is still generally frowned upon – although there are situations in which it might be tolerated, within limits. Observing a young horse taken to a competition, Latimer and Birke (2009) noted how the horse was plunging around on the end of a lead rope – behaviour which to outsiders might indicate disobedience. The human, however, completely ignored these youthful antics: to her, the horse had been brought to the showground 'for education', to see a bit of the world. The animal's unruliness was tolerated within certain limits, and seen as normal.

Riders, too, sometimes insist that the horse must learn to 'think for him/herself' during riding. This is not so much about 'waywardness', or about obedience or subservience. Rather, it is more about riders sometimes wanting to rely on horses' own abilities to work something out for themselves, to take control of situations, at least occasionally. This may be an ability to find their way home, if horse and rider find themselves lost on a trail ride. Or it may be the ability of the horse to help out during a jumping competition. Amateur riders competing in show jumping, for example, put great store on finding horses who can make their own judgements of stride or distance, and who have the ability to take charge and manage the jumping effort despite the rider's mistakes (Smart 2011; Thompson and Birke 2014). But this is more than simply a technical skill on the part of the horse; rather, riders see it as a crucial *part* of the affective relationship.

Whatever the role of the horse, horse people inevitably try to find explanations for the animal's ways of behaving. Do they see their horse as a potential partner, someone who cooperates closely with them? Do they see the horse as disobedient, behaving in ways that pose problems to the person? Or, do they see the horse as expressing his/her own personality, even when being less obviously 'cooperative'? Here, we consider these questions, drawing primarily on interviews with horse people, who were talking about individual horses' personalities and responses, as well as considering two case studies. The interviews were from our own research on horses and people, particularly from a study of horse–human relationships.[4]

In trying to explain the horse's behaviour, handlers would tell stories of how they and the horse were reacting to each other during the encounter. Sometimes, they acknowledged a sustained lack of connection, particularly if they did not know the horse beforehand; at other times, they emphasized depth of connection (notably with their own horse) – how well they had worked together. These were stories of attunement, of mutuality. But, even when they stressed connection, they also noted short moments of disconnection, which had to be explained, as we will argue.

Not 'gelling': when things are not quite right

Sometimes, it is obvious to outside observers that a person and their companion animal have a bond: they seem to enjoy each other's company, to pay attention to what the other does. At other times, however, the interaction is clearly difficult, and neither seem to be enjoying the encounter. While there may be dire consequences for the animal if humans perceive the behaviour as problematic, many owners seek ways to solve the underlying problem, to try to change the animal's behaviour. Success then depends, at least partly, on how skilled the person is at handling that species, and how well they understand what the animal is trying to tell them. Even for people experienced with horses and riding, some relationships simply do not 'gel', and the communication between horse and rider remains fraught (Smart 2011).[5]

Not 'gelling' might be understood as a problem of mutual communication. In some cases, owners might see the horse as being the problem, using labels such as 'wilful' or 'disobedient', even 'bad' or 'dangerous'. In other cases, owners might acknowledge their own failure to communicate with the horse, and seek to work on the communication, perhaps with the help of someone else. T is one example (see Case Study 1); novice handlers typically found T difficult.[6] In this case, the handler received help from an experienced behaviour consultant, who helped her to read T's signals, and to communicate with him more effectively.

In T's case, the problem was, apparently, solved through such consultation and both learned to work better with each other. But sometimes, horses have persistent issues that owners cannot (or do not want to) change – perhaps an ingrained habit[7] – despite an owner's previous efforts to ameliorate the problem. People then seek to explain away the animal's apparent recalcitrance. They might, for example, suggest that the horse had had a 'difficult life' prior to them, resulting in 'bad habits' (such as stereotypies), which cannot entirely be eradicated. Stories about the alleged previous life might include, for example, 'being on a racing yard', or, 'being in a riding school' (so not treated individually: see Birke *et al.* 2010).

Handlers involved in our study of human–horse relationships similarly sometimes invoked stories of a sorry past, for horses they knew. But those handlers who did not know the horse had to use a different strategy, centred on what the horse was doing at that moment. So, if they saw the interaction as not harmonious, they tended to refer to the horse as 'not listening', not paying attention to what they want. With an animal as big as a horse, not paying attention could potentially be dangerous, so it is a constant concern for handlers. Thus Ellen spoke during interview of the young horse, Pat, whom she had been leading, as 'completely distracted, I'm not sure I would have known he was on the other end of the rope... he wasn't engaging with me at all'. Communication is evidently lacking here, as Ellen tries to get Pat to adjust to her actions – without much success, and Pat repeatedly pulls back, resisting Ellen's attempts to lead him quietly. Kez, an older thoroughbred mare, similarly, did not communicate well with someone she did not know, and resisted frequently, being seen as 'very distracted' by the handler.

Case Study 1: T's story

T was a piebald cob, a gelding approximately 160 cm high, who was not easy to handle, as he had not had much handling and socialization when he was younger. Watching him being handled by an inexperienced handler, it was evident they both mistrusted each other, and T quickly got the upper hand – he was behaving in ways that were potentially dangerous. He would barge into his handler until she backed away, he would bite, sometimes getting hold of her sleeve, he would push ahead of her so that she was dragged along. T's body language was clear: his tail and head were elevated, tension manifest in his posture. His handler appeared, unsurprisingly, nervous around him, lacking in confidence, and she ended up frequently pushing against his shoulder in attempts to stop him barging her. She would back off, then approach, then back off, raising and lowering her hand and so altering the tension on the lead rope. Nothing was consistent, and the overall picture was unhappy; they did not work together. There was no cohesion – and certainly no gelling.

With help, however, she learned to read and anticipate T's behaviour and to try to prevent him from barging into her space. Once she had learned to work with him, he became more amenable; his body became relaxed, and they were able, at last, to walk alongside each other without a struggle, to begin to gel with each other.

Gelling: when things go well

> They are thinking animals, they will do what they choose to... they will
> certainly make decisions that will not necessarily agree with the decisions
> that we make [but] the majority of horses would like to please us... if we can
> prove to them that we are trustworthy... in most cases they will choose to be
> with us (behaviour consultant, interview[8]).
>
>   I wasn't just pulling him along, it wasn't as much a question of leadership,
> it was more a cooperation and he was more... up at my side and there was like
> we were going through it together, making, it was more like joint decision
> making... [which] made me feel quite good, and happy (interviewee Kirsty,
> after handling a horse unfamiliar to her).

Although there is always scope for misunderstanding, horse people do indeed tend to
see horses as 'choosing to be with us', as wanting 'to please us', as the first commentator
observed. When they do, then gelling becomes possible. In the second quotation,
Kirsty is talking about her feelings while she was leading a horse whom she did not
know (Rusty). Despite their unfamiliarity with each other as individuals, both knew
how to work together with the other species, to make decisions and to move together.
This, in turn, moved Kirsty emotionally, and she spoke of how the joint decision
making (apparent in this case despite not knowing each other) made her feel 'happy'.

For many horse people, making 'joint decisions' is crucial to their engagement
with their specific horse, part of the deep bond. When relationships work well, both
horse and human might experience trust, cooperation, and security – and owners
regularly tell stories of companionship and mutual affection. We turn now to three
themes horse people identified in interviews about 'relationships that work'. These
are: that because of the close bond, horses will see 'their' person as a source of
reassurance if being led away by someone else; that horses and people who know
each other well cooperate closely during moving around; and that, within a successful
relationship, slightly 'wayward' behaviour by the horse might be tolerated.

Seeking reassurance – where is my person?

Sometimes the horse we observed seemed to be following quietly, 'not having a
care in the world', apparently unfazed by the task at hand. Thus, Erica said of the
unfamiliar horse she had been leading during the exercise, that '[Rolo was] walking
in [my] shadow [...] very laid back, chilled out' (group interview). Rolo, however,
worked in an Equestrian Centre, and was handled daily by several people; most of
the other horses in our research were, by contrast, privately owned, and handled
primarily by one person, with whom they had a long-lasting bond. Not surprisingly,
perhaps, observers sometimes reported that horses led by someone else were not
unfazed, but rather seemed a little anxious, often looking at their owners 'for
reassurance': that is, the person who the horse sees every day becomes a source of
comfort and security. Thus, Kez was frequently 'turning round to look at Louis',
and clearly differentiated between him and unfamiliar handlers (as noted by Jen,

her handler at this point; see Case Study 2), while Sandy, who ran a livery (boarding) yard for horses, felt that her horse Rusty was giving her 'questioning looks' about being led by someone else. Nessa, similarly, watched her young cob, Finn, being handled by another person, and said that he 'kept looking to me for reassurance'. While this is partly wishful thinking on the part of horse owners, wanting to think of their horse behaving differently with someone else, our observations backed this up: approximately a third of the horses involved in our study checked up on (looked at) their 'owner' while they were being led by another person. What this indicates is that, although the horses are compliant, doing what is asked of them, they also do so within an affective context – their connections to 'their' particular person.

Case Study 2: Kez's story

Kez was a brown thoroughbred mare, 11 years old when we met her, and about 160 cm tall. Her 'owner', Louis, was a student, and kept Kez at the college during term. He took care of her daily, and during vacations did some novice jumping competitions with her. According to Louis's comments during interview, Kez is 'very laid back […] she is very relaxed around human contact' – and she did indeed appear to be so when he was handling her. She walked quietly alongside him, seeming to adjust her pace to his – in short, they gelled.

Once he stopped handling her, however, and went to stand away from her while we readjusted straps, and held her, then it was a different story. She seemed nervous, and unhappy to walk beside someone she did not know; she tried to pull ahead, and put her head in the air. Her handler at this point, Jen, noted that:

> She took a long time to settle down... she was very distracted by [external] noises... and she was turning round to look at Louis... she seemed very reactive and anxious... [when she was with Louis] she looked responsive, ... she looked like she was listening to him, she was waiting for him to direct her... she was watching what he was doing, whereas with me, she was just rushing through it.

The difference in her demeanour with the two people leading her was dramatic. Kez was a particularly clear example of a horse who looked for their human while being handled by someone else. That she and Louis were paying close attention to each other was very evident from our slow-motion video records. Kez's body was tense; tail elevated, and head high while we dealt with her surcingle. But Louis was behind her, some metres away, and as she raised her head, he tilted his body slightly to one side. This was a tiny movement, but enough to bring him into the edge of her field of vision and her into his. She then relaxed – and so did he, as he finished the step to one side. These two were clearly highly attuned to each other, making sure that they kept the other in view. The emotional connection was obvious.

This was particularly obvious with Kez, who apparently wanted to make sure she knew where her person was throughout. The bond was an affectionate one; Kez was clearly attached to Louis, and paid attention to where he was (as he did to her). But this was not simply that he provided care for her: rather, it is an indicator of the strength of the affective bond. The human, that is, became a 'safe haven' – a source of reassurance at moments of potential anxiety or separation (Topál and Gácsi 2012). Horses are, unlike the dogs studied by Topál and Gácsi,[9] perhaps more used to the idea that humans come and go, since horses rarely live in houses. But they can still build an attachment to specific humans – and looking often at their owners while being asked to move away from them may be testimony to that. For Kez at least, moving away from her human was indeed worrying; he was, it appeared, a source of comfort and reassurance, and her trying to keep him in view made clear the affective ties.

Working together – cooperation

The second theme was cooperation, working together. Within a close relationship, whether that is with the same or a different species, individuals can sometimes seem to move together, in ways born of predictability and familiarity. In horse–human relationships, for example, both partners can coordinate with each other, and display a kind of dynamic mutuality – whether that is during riding, or when working together on the ground (see Argent 2012; Birke and Hockenhull 2015). This was totally absent for T, before intervention, but was very evident in the case of Kez and Louis.

Many of our interviewees commented on this togetherness, referring to the coordination of horse and human.[10] Louis, for example, noted the closeness he had with Kez, saying that 'We've got quite a one-to-one relationship, that, we are [...] together most of the day and she's, just, relaxed around me [...] she'll just follow me round the school and I leave her to it', while another rider, Rachel, spoke of her horse, Bill, as being 'quite logged onto me [...] he just follows me'. She and Bill were evidently relaxed together, and indeed appeared 'logged on' to one another.[11]

Horses and humans who know each other well pay attention to one another, producing an evident mutuality; they are visibly more relaxed. In that sense, the moving together is a consequence of the relationship, an acting-out of an affective history together. But that togetherness is also a way of negotiating and (re)producing the relationship. Within that bond, affect thus both produces and emerges from the act of moving together. Trust is central – and, indeed, several interviewees noted how much they trusted their horse.

Some horses were clearly trusting; others seemed less so. Some apparently 'hadn't a care in the world', while others seemed to want reassuring. Some were described as cooperative, at least during the task we asked them to do. But another striking theme emerged from the interviews: asked to talk about their horses' personalities, many owners described their horse as occasionally being rather less cooperative, as being 'cheeky' – a term which implied tolerance of otherwise disobedient behaviour.

Tolerating 'cheekiness'

The 'cheekiness' of the horse was identified by several owners, who were also quick to label their bond with their horse as 'strong'. These narratives implied tolerance, a willingness to forgive or ignore the horse's wayward behaviour (at least within limits). This was different from not being able to communicate (as in T's case): on the contrary, all the people we observed said they were (and appeared to be) confident around horses, and all felt they had a strong bond with their animal. They could work well together – yet also acknowledged that, at times, they would put up with the horse 'misbehaving', to a degree.

Two thirds of the horse owners volunteered that they thought their horse was inclined to be 'cheeky', 'opinionated' or 'forceful', or perhaps 'playful' – 'making me laugh'. One example of this was Mel's description of her horse, as 'very cheeky... he likes making you laugh... the other day he... did a canter pirouette just so he could reach his bum to the wings [of the jump] and knock it over... he plays with poles and gets three legs on and then dangles the other leg off'. Here, she describes his wilfulness, deliberately going over to knock over an obstacle – actions which she clearly tolerated, even condoned. Commenting further on how he walked through some obstacles during the research, she said, 'it made me laugh because he's wanting to stop and talk to people... he wasn't being naughty, it was just... I thought he was going to play [with the poles]'. In this example, Mel is acknowledging that the horse was expected to walk alongside her without stopping, though she permitted him to stop and 'talk to people', and even laughed at his antics; yet to her, these were not manifestations of 'naughtiness', but expressions of his personality.

Another horsewoman, Babs, spoke similarly about her horse, Ben: 'he's very cheeky... He likes to know what's going on... [but] when I'm grooming him, he will bite me on the bum and he will pull at my clothes... he loves attention a lot, Ben does, but, he's just cheeky in ways'. Biting is hardly a desirable behaviour in a companion animal, but as long as it is a nip, pulling at clothes, Babs seems willing to tolerate it, as part of her relationship with him, as part of his cheeky personality.

Cheekiness seems to be explicitly allowed by these horse owners. Rachel, for example, spoke of her horse, Bill, as 'cheeky, and naughty, but he never does anything that would hurt you, whether that's intentional I don't know'. Bill is an older horse (17 years), who has been trained to do advanced dressage, and still competes with Rachel. She goes on to explain that

> probably because he's older, he can get away with a few things... He didn't have much of a personality when I got him... he had a very, very rigorous upbringing [before], which is why I let him get away with a few things... that he shouldn't do, when we're working... He's allowed to be naughty at times, as long as he's not dangerous naughty... just allowed to be less disciplined, I think he deserves to be, he gets away with a few things.

In this case, Rachel makes reference to his past, which she sees as over-disciplined, and explains his cheekiness in terms of her making allowances for his background.

Observing her work with Bill, there is an evident affectionate bond. Rachel talks about how he is allowed to wander around the yard at home, 'sticking his head in the kitchen window' if he wants to say hello: but he is expected to 'work properly, he has to work for his living' when he is being trained for dressage. Some degree of cheekiness is thus tolerated, but within limits: the horse is expected to understand where those limits lie.

We want to emphasize two things emerging from this discourse of cheekiness. First, these horses are clearly expressing themselves, doing things which – to some people and in some contexts – would be considered undesirable, even 'naughty', or a problem. This, then, raises an important question: to what extent will owners tolerate occasional waywardness? When do owners find their animal's behaviour to be intolerable, and when do they simply ignore it? We saw with T that nipping was part of the problem – he was trying to dominate his handler, and using his teeth to make the point. In no way were they working together. But by contrast, Babs is willing to ignore occasional nips from Ben; to her, nips are not a problem because on the whole they gel and Ben is willing to work with her, at least during our observation! How and when behaviour is deemed a 'problem' by people is an important issue, with considerable implications for horses' welfare.

Second, while there is an evident cohesiveness in the way that familiar horses and people (in an established relationship) move together, the horse can break out of this at times, and move away, from the person, from the desired direction of movement. Yet this moving away playfully has a reference point, as does moving away with a strange person – and that reference is the affective bond with the familiar other. For all that the horse is asserting him or herself, it is within an affective context. But sometimes, an animal's behaviour exceeds those arbitrary limits, potentially becoming a problem.

Discussion: connections alongside

Such wayward behaviour, slightly pushing the boundaries of what is acceptable to humans, was not seen as particularly problematic for the people we interviewed in our research, most of whom defined themselves as reasonably experienced and confident around horses. But horses sometimes behave in ways that people cannot accept, for whatever reason. Owners can then either choose to relinquish (usually by selling) the horse, or they may seek external help, through a trainer, a vet, or a specialized behaviour consultant. Owners taking this route usually do so not only to solve the problem, but also because they seek a better relationship.[12] How and when people define the horse's behaviour as 'problematic' are questions that need more research, though they clearly have profound consequences for horses. There is a world of difference between labelling horses as 'bad' and labelling them as 'cheeky'.

Despite affective ties, horse–human relationships are often broken, in part because of the monetary value of horses. When we finished our observations of Kez, Louis mentioned that, for all that bond, and attachment, he was going to sell her. Humans expect and want horses to make close, unique bonds with us; indeed, horses' ability to do this seems to have a big impact on whether we gel with them or not, whether they stay or are sold on. Yet we can still break the bond as we see fit: we may be moved by them, but we can move them on, without knowing – or perhaps even caring – about their emotional response. And, once they are rehomed, what was one person's cheekiness becomes another person's behaviour problem.

Many owners, however, do tolerate at least some potential 'problems', within a close relationship, and cheekiness seemed to be accepted among our respondents as the horses' way of expressing themselves. But it is expression within limits; pushing poles around, nipping clothing or simply being less disciplined, are tolerated provided they do not threaten the overall ways of the horse's engagement with the human. On the contrary, they seem to be seen as somehow contributing to the relationship. Within affective ties, it becomes possible for the horse to move away, even if partially and temporarily, to assert her/his desires.

The cheeky horse is breaking out of bounds, erring toward the wild, in ways that for confident handlers may be laughingly tolerated. Horse people often acknowledge that their animal is 'feeling a bit fresh today' (that is, being a bit wild, perhaps bucking or running off – responses which may be terrifying to a nervous or novice person). Nevertheless, that people often put up with such responses suggests that absolute 'obedience' is not completely desirable – and it is noteworthy that none of the respondents in our research projects have ever spoken of 'domination' of their horse, but have consistently emphasized cooperation and the mindedness of their horse, although several also mentioned the need for 'respect'. Many riders, on the contrary, put a premium on horses who have 'minds of their own', which they can bring to the relationship – to help out, occasionally, as we noted earlier.

Within human–animal studies literature, there has been considerable emphasis on mutuality between us and other animals. What we observed as horses and their people who moved around in harmony was an intercorporeality, albeit a partial one (see also Argent 2012; Maurstad *et al.* 2013). Just as Laurier, Maze, and Lundin (2006: 2) speak of dog-walking as a 'living accomplishment' of both humans and dogs, so the cohesion of horses and people walking together is co-produced. This mutual production happens at two levels. From moment to moment, as they walk together, they pay heed to one another, making and remaking the connection. But this movement together is also a product of a shared history, of entwined biographies, and of mutual affect. Ways of gelling are ways of moving – be that spatially or affectively.

People and horses thus build expectations of one another, so that 'distraction', not moving in cohesion, or being asked to move away from the familiar, are all things which might cause concern for one or other. Such close, corporeal, working with another across species divides is clear in many cases of assistance animals, who must learn to anticipate closely the physical needs and movements of their human. Higgin (2012: 82–3), in his study of guide dogs and their people, notes how

> both co-constitute the motifs of perception and action that grow within the daily performances of making mistakes, holding back, and walking in rhythm [...] This rhythmical understanding of the body [...] is neither objective nor subjective: it is affective – knowledge here is the ability to affect and be affected by others.

Affect, in other words, is produced and experienced relationally and intercorporeally, through experience of, and engagement with, each other, whether that be within or between species (Blackman and Venn 2010).

In Haraway's (2003; 2008) influential work on human–animal relationalities, she stresses the interconnectedness of our lives with 'companion species', suggesting (2003: 6) that '[b]eings do not preexist their relatings'. Rather, being-with, in relationality, is a cornerstone of social life, and her work reminds us forcefully of how we are profoundly connected, whatever our species. In emphasizing the sense of horse and human moving together described and observed in our research (Birke and Hockenhull 2015), we focus on such interspecies relationality. In many ways, our respondents' narratives evoked the image of the mythical centaur (see Game 2001; Thompson 2011) – a hybrid figure, half human, half horse, which symbolizes the deep, intercorporeal, connection to which horse people aspire.

Yet connections are not continuous. For all that horse people wistfully refer to feelings of 'oneness', of almost 'becoming horse' (Smith 2011), there is simultaneously a recognition in interview narratives that the horse can also separate out, that s/he remains an/other, who can be distracted, or – at times – permitted to push the boundaries, to remain 'other'. While broadly sympathetic to Haraway's line of argument, Latimer (2013) takes issue with the centrality of the idea of 'being-with' in much theorizing about humans and animals, which for her valorizes and reifies hybridity (albeit partial). However useful the idea of hybridity is, it risks, she suggests, becoming totalizing, invoking a 'dyadic hybrid, whose parts become something other than they are without becoming with the other. And […] it is about being infolded, as an inward, not an outward movement' (Latimer 2013: 92–3). Rather, she prefers the idea of 'being-alongside', which carries a sense of partial connection/partial division, and of moving outward from the other (Latimer 2013). Relations of humans and non-human animals entail, she insists, partiality and intermittency, and it is these partial – and overlapping – connections which produce social life. Thus, however close we may feel that bonds are with others as we move together, we, and they, also move apart.

It is this sense of partial connection that, we would suggest, is implicit in the way that owners so often want to explain away their animal's occasional waywardness, within an otherwise working relationship. Our respondents, all reasonably confident with horses, and with shared histories with their own horse, both permitted and explained what the horse did on these occasions. They wanted to prioritize the bond with the animal, but they also recognized the partial connection, expressed by the horse's movements away from them, or away from what humans desired. Talking about horses' cheekiness was usually accompanied by wry smiles, as though the behaviour was not merely tolerated but was seen to be part of the way that horse and human are with each other. In that sense, telling stories of a horse's wayward behaviour could be seen as helping to configure the relationship. The relationship that is thus portrayed consists not only of moments of profound connection, but also of disconnection, while the horse expresses his or her unique personality.

Perhaps we cannot know the horse's experience of such being-alongside, although we can try to acknowledge their part in building the relationship, and to consider animals' emotional responses (see also Bekoff 2002; Fraser 2009; Wemelsfelder 2012). Relationships that work, be that with other humans or with other species, entail working together but also imply tolerance. The riders in our study permitted their horse some leeway, some forms of expression, which they were willing to tolerate – within limits. Perhaps the horses, too, were sometimes willing to ignore some of their human's foibles.

Partial connection and separation helps to produce different biographies – of each partner, as well as that of the relationship. The various stories people tell of their horses' personalities (and the way that they are with their horse) enact differing versions of the human–horse relationship.[13] To portray horses as helpers is to imply that they are competent participants in shared activity – partners with minds of their own; to describe how one might tolerate horses' antics when they are being cheeky hints at the potential 'wildness' or self-expressivity of horses, their capacity to pull away from 'obedience'. These stories, however, coexist with the trope of the horse as 'friend', a narrative which enacts a tale of close companionship and mutual affect, as well as tolerance. Both mutuality and separating out create multiple ways of telling stories about our connections with non-human animals.

'To know someone […] is to know their story, and to be able to join that story to one's own' (Ingold 2011: 160–1). Journeys together – whether with horses or another species – are not just about human and animal working in harmony (though they are partly that). They are also about separating and coming together, about our being in the world, and about sharing stories along life's paths. They are about moving, and being moved, about being affected. Maybe horses cannot directly tell us their stories: but if we observe and interact with them closely, then perhaps we can hope to join their stories with our own.

Acknowledgement

We are very grateful to Felicity George for discussions about communication in horse–human interactions, for sharing her ideas with us about what happens when 'things go wrong', and for helpful comments on the manuscript. We also want to thank Tami Young for discussions about equine welfare, and for taking part in the ongoing research.

Notes

1 While we acknowledge that the idea of animals as possessions (hence, having 'owners') is deeply problematic, the term 'owners' is widely used. Moreover, since horses are typically bought and sold, sometimes for large sums of money, it is particularly appropriate for this species.

2 If relationships do not work out, for whatever reason, animals are likely to be relinquished – to rehoming centres for example – or euthanased (see Mondelli *et al.* 2004; Shore 2005; Crowell-Davies 2008).

3 We recall a humorous commentary a few years ago, in the British equestrian magazine, *Horse and Hound*, which claimed that a survey of their women readers revealed that 70 per cent preferred to spend time with their horse than with their husband.

4 For that particular study, we conducted short interviews with participants in the study of horse–human relationships (N = 30 interviews, 21 with familiar partnerships and nine with unfamiliar ones), as well as one focus group of people working with handling unfamiliar horses. For further details, see Birke and Hockenhull (2015). All interviewees' names and the names of their horses cited here have been anonymized. Although the horses involved in our research studies are of both genders, the humans are almost entirely women – in keeping with the demographics of leisure horse riding in general.

5 Indeed, 'not gelling' is a reason often given for selling the horse on (see Heijtel 2012). It is also a reason for horse owners to seek help from professional behaviour consultants (F. George, personal communication, March 2014).

6 Felicity George (personal communication, March 2014) notes that owners having 'problems' with their horse do sometimes describe the horse's personality in negative terms, such as 'disobedient'. George kindly supplied the video footage on which the description in Box 10.1 is based. The video can be seen on www.youtube.com/watch?v=vB4ZcD9H1Yc, (accessed March 2015). It shows the poor communication between T and his handler before, and after, George had advised on how to work with T. The difference is considerable.

7 Some behavioural responses develop into stereotypies, which are extremely difficult to eradicate. Among horses, the most common are 'crib biting' (repetitive chewing on, e.g., the stable door), 'wind sucking' (a further development which involves sucking in air), and 'weaving' (side to side repetitive movements). These are typically associated with the restrictions imposed by many methods of husbandry (e.g. keeping horses in individual stables).

8 This is a separate, ethnographic, study of horse people seeking help from outside experts – notably, behaviour consultants (Birke and Miele, unpublished).

9 Topál and Gásci (2012) studied dogs and their responses to owners leaving the room. They interpreted the dogs' behaviour in terms of attachment theory, originally developed in relation to human children by John Bowlby.

10 This was corroborated by our field notes, as well as an independent panel of observers (see Birke and Hockenhull 2015).

11 Indeed, their heart rates were correlated throughout the observation period (unpublished data; see also Hockenhull *et al.* forthcoming).

12 There are many people offering services as behaviour consultants, aiming to help people solve behavioural problems. The backgrounds and skills of consultants are diverse, and there is currently little accreditation or validation, at least in the UK. Owners seek help for many reasons, but part of it is wanting to have a better relationship with the animal. One consultant estimates that approximately 60 per cent of clients emphasize that they want the 'horse to like me', 'just want him to love me' (F. George, personal communication, March 2014).

13 This point draws on Mol's (2012) discussion of dieting techniques; she suggests that consumers are faced with innumberable choices about nutrition and food, so that choosing one dieting technique over another in effect *enacts* different versions of food, and incorporates different norms.

References

Argent, G., 2012. Toward a privileging of the nonverbal: communication, corporeal synchrony, and transcendence in humans and horses. *In:* J.A. Smith and R.W. Mitchell, eds. *Experiencing animal minds: an anthology of animal–human encounters*. New York: Columbia University Press, 111–28.

Arluke, A., 1994. We build a better beagle: fantastic creatures in lab animal ads. *Qualitative Sociology*, 17, 143–58.

Bekoff, M., 2002. *Minding animals: awareness, emotions and heart*. Oxford, UK: Oxford University Press.

Birke, L., 2007. 'Learning to speak horse': the culture of 'natural horsemanship'. *Society and Animals*, 15, 217–40.

Birke, L., 2008. Talking about horses: control and freedom in the world of 'natural horsemanship'. *Society and Animals*, 16, 107–26.

Birke, L. and Hockenhull, J., 2015. Journeys together: horses and humans in partnership. *Society and Animals*, 23, 81–100.

Birke, L., Hockenhull, J., and Creighton, E., 2010. The horse's tale: narratives of caring for/about horses. *Society and Animals*, 18, 331–47.

Birke, L. and Miele, M., unpublished. How can I live well with my horse? The newcomers to the 'horse world' and the challenge to horse training.

Blackman, L. and Venn, C., 2010. Affect. *Body and Society*, 16 (1), 7–28.

Crowell-Davies, S., 2008. Motivation for pet ownership and its relevance to behavior problems. *Compendium on Continuing Education for the Practising Veterinarian*, 30 (8), 423–8.

Cudworth, E., 2011. *Social lives with other animals: tales of sex, death and love*. Basingstoke, UK: Palgrave Macmillan.

Fraser, D., 2009. Animal behaviour, animal welfare and the scientific study of affect. *Applied Animal Behaviour Science*, 118, 108–17.

Game, A., 2001. Riding: embodying the centaur. *Body and Society*, 7, 1–12.

Haraway, D., 2003. *The companion species manifesto: dogs, people and significant otherness*. Chicago, US: Prickly Paradigm Press.

Haraway, D., 2008. *When species meet*. Minneapolis, US: University of Minnesota Press.

Heijtel, M.G., 2012. Movement of horses between owners in Great Britain. Unpublished dissertation thesis. University of Bristol, UK.

Higgin, M., 2012. On being guided by dogs. *In:* L. Birke and J. Hockenhull, eds. *Crossing boundaries: investigating human–animal interactions*. Leiden, Netherlands: Brill, 73–88.

Hockenhull, J. and Creighton, E., 2013. Training horses: positive reinforcement, positive punishment, and ridden behavior problems. *Journal of Veterinary Behavior*, 8, 245–52.

Hockenhull, J., Young, T., Redgate, S., and Birke, L., forthcoming. Two hearts that beat as one? Exploring differences and synchronicity in the heart rates of familiar and unfamiliar pairs of horses and humans completing an in-hand task. *Anthrozoös*.

Ingold, T., 2011. *Being alive: essays on movement, knowledge and description*. London: Routledge.

Latimer, J., 2013. Being alongside: rethinking relations amongst different kinds. *Theory, Culture and Society*, 30 (7/8), 77–104.

Latimer, J. and Birke, L., 2009. Natural relations: horses, knowledge and technology. *The Sociological Review*, 57 (1), 1–27.

Laurier, E., Maze, R., and Lundin, J., 2006. Putting the dog back in the park: animal and human mind-in-action. *Mind, Culture and Activity*, 13 (1), 2–24.

Leckie, E.J., 2001. *The equine population of the UK*. Norfolk, UK: International League for the Protection of Horses.

Maurstad, A., Davis, D., and Cowles, S., 2013. Co-being and intra-action in horse–human relationships: a multi-species ethnography of be(com)ing human and be(com)ing horse. *Social Anthropology*, 21, 322–35.

McElroy, S.C., 1996. *Animals as teachers and healers: true stories and reflections*. Troutdale, US: New Sage Press.

Mol, A.-M., 2012. Mind your plate! The ontonorms of Dutch dieting. *Social Studies of Science,* 43, 379–96.

Mondelli, F., Previde, E.P., Verga, M., Levi, D., Magistrelli, S., and Valsecchi, P., 2004. The bond that never developed: adoption and relinquishment of dogs in a rescue shelter. *Journal of Applied Animal Welfare Science,* 15, 253–66.

Oxford English Dictionary, 2015. Online ed. Available from: www.oed.com [Accessed 24 March 2015).

Proops, L. and McComb, K., 2010. Attributing attention: the use of human-given cues by domestic horses (Equus caballus). *Animal Cognition,* 13, 197–205.

Schwab, C. and Huber, L., 2006. Obey or not obey? Dogs (*Canis familiaris*) behave differently in response to attentional states of their owners. *Journal of Comparative Psychology,* 120, 169–75.

Shore, E.R., 2005. Returning a recently adopted companion animal: adopters' reasons for and reactions to the failed adoption experience. *Journal of Applied Animal Welfare Science,* 8, 187–98.

Smart, C., 2011. Ways of knowing: crossing species boundaries. *Methodological Innovations Online* [online], 6 (3), 27-38. Available from: www.pbs.plym.ac.uk/mi/pdf/8-02-12/MIO63Paper21.pdf [Accessed 20 July 2015].

Smith, S.J., 2011. Becoming horse in the duration of the moment: the trainer's challenge. *Phenomenology and Practice,* 5 (1), 7–26.

Smuts, B., 2006. Between species: science and subjectivity. *Configurations,* 14, 115–26.

Thompson, K., 2011. Theorising rider–horse relations: an ethnographic illustration of the centaur metaphor in the Spanish bullfight. *In:* N. Taylor and T. Signal, eds. *Theorizing animals: re-thinking humanimal relations.* Leiden, Netherlands: Brill, 221–53.

Thompson, K. and Birke, L., 2014. 'The horse has got to want to help': human–animal habituses and networks of relationality in amateur show jumping. *In:* J. Gillett and M. Gilbert, eds. *Sport, animals and society.* New York: Routledge, 69–84.

Topál, J. and Gácsi, M., 2012. Lessons we should learn from our unique relationship with dogs: an ethological approach. *In:* L. Birke and J. Hockenhull, eds. *Crossing boundaries: investigating human–animal relationships.* Leiden, Netherlands: Brill, 161–86.

Tuan, Y., 1984. *Dominance and affection: the making of pets.* New Haven, US: Yale University Press.

van Kerkhove, W., 2004. A fresh look at the wolf-pack theory of companion-animal dog social behavior. *Journal of Applied Animal Welfare Science,* 7, 279–85.

Wemelsfelder, F., 2012. A Science of friendly pigs ... carving out a conceptual space for addressing animals as sentient beings. *In:* L. Birke and J. Hockenhull, eds. *Crossing boundaries: investigating human–animal relationships.* Leiden, Netherlands: Brill, 223–49.

11 Companionable human–animal relationality

A reading of a Buddhist jātaka (rebirth) tale

Teuvo Laitila

A story from the northeast Indian Naga says that tigers and humans have common origins and are thus siblings, or brothers, as the Naga say (Green 2006: 7). Similar stories are known from other parts of the Asian continent (Van Deusen 2001). In the postmodern Western world, stories of werewolves have given rise to a version of therianthropy, or spirituality lived out mainly in cyberspace, where identities between humans and (other) animals are blurred (see Robertson 2013). Of course, scientifically speaking this is not convincing; nor is the commonality between tigers and humans meant to be a piece of scholarly information but an expression of the Naga and other peoples' understanding of human–animal relations.

Considering the rapid growth of human–animal research, Lynda Birke and Jo Hockenhull note (2012: 15) that 'investigating how specific [human–animal] relationships form and are maintained' has been curiously lacking in the past. They evidently have in mind concrete relationships. These stories of human–animal relationships are worth considering, however, especially if they have some paradigmatic and inspirational value for their readers, listeners or, in film or theatre, viewers, as the *jātaka* tales have for many Buddhists (Ohnuma 2005: 102–3; Appleton 2010: 2; Kemmerer 2012: 114). Some scholars have even called the jātakas one of the three 'lineages of the Buddha', emphasizing the accumulation of the Buddha's virtues, particularly generosity, during his numerous rebirths. The two other lineages are that of the historical Buddha, Śākyamuni, the Enlightened One, and that of bodhisattas, buddhas-to-be, appearing from aeon to aeon to redeem all sentient beings (Gombrich 1977: xvii; Strong 2011: 171, 181). In this chapter I focus on how one jātaka, known as 'The Tale of the Tigress' (see Khoroche 2006: 5–9), may be interpreted from the viewpoint of human–animal relationality. I first introduce the genre and the role of animals in it, then summarize the tale, and finally suggest an interpretation.

Jātaka tales

According to the *Oxford Dictionary of Buddhism*, the Pāli-language word jātaka refers to a 'genre of early literature describing the former lives of Gautama Buddha' (Keown 2004: 125). The term comes from the Sanskrit *jāti* meaning birth, indicating a chain of beings culminating in Śākyamuni. In Theravāda

Buddhist scriptures, 547, generally, short stories about the Buddha's previous births are included in the tenth section of the Khuddaka Nikāya or Minor Collection, which in turn is part of the Sutta Pitaka, the Book of Learning,[1] dating in written form to the first century BCE. The jātakas were taken over, and reworked, by Mahāyāna Buddhism (Gombrich 1977: xxvii–xli; Keown 2004: 125; Appleton 2010: 7–19, 66).

The exact history of the formation and compilation of the jātakas is disputed. They seem to have existed as a distinct genre in the second century BCE, at the latest. The earliest tales were evidently adapted from the ancient Indian moral stories with animal characters, later collected in a work entitled Pañcatantra (Five Principles) (Kemmerer 2012: 94, 113–14), or included in the Hindu epic *Mahābharata*. After the Buddha's passing away ca 400 BCE, jātakas were told to describe his 'evolution' during his earlier rebirths, which he was said to have remembered at the moment of his enlightenment. The tales were used in the monastic and general education of Buddhist virtues, particularly generosity, compassion, modesty, and respect of all sentient beings, and were revised for their present form during the first centuries CE by Buddhist men of letters. Today, jātakas are still used for educating people, particularly children, in Buddhist ethics and doctrine (Appleton 2010: 41–64, 78–83).

The specific tale I have focussed on here is attributed to the otherwise unknown Ārya (Reverend) Śūra, evidently an erudite Mahāyāna monk and compiler of the Sanskrit-language work Jātakamālā (Garland of Birth Stories). The text dates from the fourth or early fifth century CE and consists of 34 tales, mainly based originally on Pāli-language stories (Gombrich 1977: xxxi, xxxvi; Khoroche 2006).

Human–animal relations in the jātakas

In theory, all Buddhist schools emphasize a benevolent attitude towards nature in general and animals in particular. They base their view on various points. One is a statement ascribed to the Buddha soon after his birth: 'For enlightenment I was born, for the good of all that lives' (Kemmerer 2012: 93; see also Chapple 1997: 133–4, 138–140; Harris 2006: 207–10). The other points include the three central aspects of Buddhism as it is commonly practiced: *ahimsā*, not harming any living being, *mettā*, loving-kindness or unselfish intentions and the capacity to make the others happy, and *karuna*, compassion, or the intention and capacity to relieve the other's suffering (Harris 2006: 211; Kemmerer 2012: 95–9).

From a strict Buddhist view, jātakas, as well as other Buddhist texts dealing with animals, are really not about humans and human–animal relations but, rather, are about doctrinal metaphors and relations between humans (see Deleanu 2000: 81; Harris 2006: 208). Canonical Buddhist traditions stress that it was in the human form that the Buddha was enlightened. While jātakas do not stress this, they do often present him in an animal form, indicating that he was pledged to redeem all sentient beings. In the 547 Khuddaka Nikāya jātakas 70 different types of animals are mentioned, only some of which are the Buddha. The most common are monkeys (in 27 stories), elephants (24), jackals (20), lions (19), crows (17), and deer (15).

The tales present the animal-Buddha as an exemplary being. In the form of a water buffalo, he patiently tolerates the harassing monkey. As a monkey, he sacrifices his life by using his body to form a bridge to create an escape route for others; in the process his back is broken (Chapple 1997: 134; Kemmerer 2012: 94–5).

Moreover, in Tibetan jātakas, animals other than the animal-form Buddha are credited with Buddhist virtues. For example, some birds 'can easily learn dharma [Buddhist teaching]' and gain enough merit 'to attain release without rebirth' (Kemmerer 2012: 118; see also Vargas 2006: 221–31).[2] It is worth taking into account these popular traditions as well, the more so because in Buddhism the whole human–animal division is unimportant and even misleading. There are no independent 'human' and 'animal' persons or selves, merely variations of external forms originating from the interplay of continuously changing physical and mental aggregates (in Sanskrit, *skandha*) (cf. Kemmerer 2012: 103–4). Thus it is possible to read jātakas as tales on reversible relations between humans and (other) animals and ask how humans appear after such a reading (cf. Vargas 2006).

The tigress and the Bodhisatta[3]

The jātaka under scrutiny here, the 'Vyāghrījātaka' or 'Tale of the Tigress', is not found in Pāli-language collections but is extant in Sanskrit, Chinese, Tibetan, and some other northern Buddhist languages, and is also known in vernacular in countries that practice Theravāda Buddhism (Khoroche 2006: 5–9; Matsumura 2010: 1164–5; Appleton 2010: 4–5, 119). The 'Tale of the Tigress' can be traced back to one of the most popular Mahāyāna texts, the Suvarnaprabhāsa Sūtra[4] or Sūtra of Golden Light, compiled during the first centuries CE in Sanskrit and translated into Chinese a few centuries later. The sūtra's main message is that in the universe there are four guardian devas (divinities) protecting rulers who govern their countries in the proper manner. This theme of a virtuous ruler is typical of early Buddhist literature in general (see Vargas 2006), and one may assume that originally such texts were intended to promulgate true Buddhist virtues.

In its present form, the 'Tale of the Tigress', as with jātakas in general, may be structured in three parts. These are i) a narrative set in the speaker's present, setting out the storyteller's (in most cases, the Buddha's) reasons for relating it, ii) a story of past events, or the jātaka proper, and iii) the return to the present where the storyteller points out the instructions embedded in the tale (Appleton 2010: 6, 61).

According to the tale, in one of his lives, the Bodhisatta was born to an esteemed Brahmin family. He was greatly respected and had an outstanding reputation, but was not satisfied with worldly pleasures. He therefore retired to a forest where his kindly presence had a calming effect on the wild beasts; they stopped preying each other and began to live like hermits (Khoroche 2006: 6). He soon became famous, and a multitude of (human) disciples flocked to him.

Once the Bodhisatta and his disciple Ajita saw in a mountain cave a tigress in the pangs of giving birth. The cubs were evidently born, although this is not mentioned, because the text continues: 'Her eyes were hollow with hunger, her belly horribly thin, and she looked upon her whelps [...] as so much meat'

(Khoroche 2006: 6–7). The Buddha-to-be took this to indicate that the instinct for self-preservation is stronger than a mother's affection to her offspring (Khoroche 2006: 7). This evidently shocked him, as he sent Ajita to find something to appease her hunger, promising meanwhile 'to stop her from resorting to violence' (ibid.).

Sending Ajita away was a pretext, for the Bodhisatta thought: 'Why search for meat from some other creature when there is my entire body available right here?' (Khoroche 2006: 7) He continued by stating that for him the body was useless, 'a miserable ungrateful thing' (ibid.). One would therefore be 'a fool not to welcome the chance of its being useful for someone else' (ibid.). Moreover, he claimed that any other kind of behaviour would be selfish, and considered feeding the tigress as an example to all who worked for the good of the world, 'an encouragement to those who falter, a delight to those who are practiced in charity, a powerful attraction to noble hearts' (Khoroche 2006: 8). He went on to state that he could not be happy 'as long as there is someone who is unhappy' (ibid.), and mused that his act of self-sacrifice would be an example to all who strive for the good of the world. Then he hurled himself down into the tigress's cave, just at the moment when she was about to devour her young. In Buddhist terms, he offered himself to preserve the principle of *ahimsā* by not letting the mother eat her own babies, and in order to vindicate the power of affection over violence.

Ajita, meanwhile, returned without having found any meat. After he perceived what had happened, his feeling of grief was countered by amazement at such an extraordinary deed. He said: 'Oh! How compassionate the Noble One has shown himself to those in distress [...]. Oh! What supreme love he has shown – bold, fearless, and full of goodness' (Khoroche 2006: 9). This does not mean that the *act* of giving was important as such. What mattered was the Bodhisatta's *intention* of giving; his selfless generosity meant that he was unattached to the gift of his body (Gombrich 1977: xxii–xxiii).

In the Sūtra of Golden Light, as well as in Tibetan-origin variants of the tale, the gift-giver is called Prince Mahāsattva (Great Being, another title used for the Bodhisatta); hence these versions are known as the Mahāsattva Jātaka. The main plot is the same: the prince encounters a hungry tigress ready to devour her two cubs. He slits his throat (or otherwise hurts himself), allowing the tigress to drink his blood and, finally, to eat him. In this tale, the cubs are two thieves that the Buddha, in a previous life, had saved from execution (Ohnuma 1998: 329–30). This variant emphasizes the Bodhisatta's generosity more to humans than the animals. In two 'reverse' jātakas, a rabbit and an elephant, both animal forms of the Buddha-to-be, offer their bodies to starving humans. 'The rabbit flings himself into a fire to be cooked, while the elephant runs off a cliff at the feet of those in need' (Kemmerer 2012: 109). However, self-immolations are rather atypical to jātakas. In Khuddaka Nikāya there are merely seven such tales, or less than one per cent of all (Sheravanichkul 2008: 769–70).

Assuming that most people are ready to admit of humans being capable of generosity, with some even attributing this altruism to other animals, why, then, does Ajita find the Noble One's act so astonishing? My view is that with his exclamation, he both affirms the absolute virtue of the Bodhisatta's offer and exhorts himself, and everyone else, to follow the Bodhisatta's example of living in a truly generous manner, even if offering one's body might not seem feasible at first sight. We will return to this issue below.

Interpreting the tale

Radical Buddhist interpretation of the tale takes it as an instruction to sacrifice oneself for the buddhahood (Sheravanichkul 2008: 774). In a standard Buddhist interpretation, the jātaka reveals a selfless act of a true bodhisatta: offering one's life for another's sake is the highest of a bodhisatta's five great sacrifices (in Pāli, *mahāpariccāga*). As the Buddhist scholar Arthid Sheravanichkul puts it (2008: 780), jātakas 'in which the Bodhisatta sacrifices his body are clearly used to demonstrate the perfect example of giving [generosity]'.

The opening paragraph of the tale confirms this by stating, that 'the Lord [Buddha] [...] blessed mankind with a stream of kindnesses: he gave generously, spoke lovingly, and instigated goodness' (Khoroche 2006: 5; cf. Appleton 2010: 102). Due to the tale's emphasis on giving, Reiko Ohnuma, one of the few scholars having examined it, has called it a gift-of-the-body story. According to her, by giving his body to the tigress, the Buddha-to-be also gave her his supreme teaching, or dharma, which in the Buddhist view is the best of all gifts (Ohnuma 1998: 326; Ohnuma 2005: 104). It does not represent a mere giving but, as the Bodhisatta in the tale emphasizes, a conscious rejection of selfish conduct and a choice to live life in a selfless manner (Khoroche 2006: 7–8).

Giving a gift, or showing generosity and selflessness, is both an attitude and an act. More precisely, it is a social act, which presupposes a social context, something that Birke and Hockenhull (2012: 16–20, *et passim*) find lacking in human–animal studies. Accordingly, both Buddhist interpretations and those of Ohnuma (1998; 2005) state that although the Bodhisatta's act of giving happened in the wilderness, the act was not intended to be left behind but to be adhered to by the Buddha's followers. In other words, what the Buddha-to-be had done was a repeatable social practice. By giving his body – that is, by giving everything – the Bodhisatta saved several other beings: the tigress and her cubs, and, in the process, also himself; the latter because of his intention not to win anything for himself. In Buddhist terms, by his absolute generosity he secured a better rebirth.

But what about the tale's inconsistencies with canonical Buddhist ethics? By feeding the tigress with his own body, the Bodhisatta committed a suicide and thus violated the cardinal virtue of *ahimsā*, whereas the tigress, by eating meat, merely followed her 'natural' instinct of self-preservation; she could not be rebuked of violating ethical rules. Does this mean that the human–animal relationship is always ethically asymmetrical, requiring one of the partners to break the rules of ethics? I do not think so, but I understand that the tale suggests that there is no code of ethics common to all beings; in an ethically 'fair' relationship, each side should behave according to his/her/its prevalent form and 'nature'. The Bodhisatta 'has to be' compassionate and generous, whereas a tigress is expected to eat whatever she can hunt down. I do not suggest that ethics acts as a straitjacket upon human–

animal relations, however. There are always chances to strive for improving oneself, as suggested by a Mahāyāna jātaka that tells of how a small quail refused to 'eat worms or insects, subsisting only on the grass seeds that his mother [brought] to the nest' (Kemmerer 2012: 108; cf. Chapple 1997: 137–8). The bird's virtue was of course that of the Buddhist monks; and the tale suggests that growing up in virtuosity is not restricted to some particular form.

The tale and human–animal relationality

The jātakas presuppose two things that scholars of real life human–animal relationships have found hard to explain, namely, that interspecies communication and understanding of the other's way of experiencing the world is possible (cf. Birke and Hockenhull 2012: 15–16). 'The Tale of the Tigress', too, suggests this and indicates that distinctions such as 'human–animal' or 'material–spiritual' are perceived, not real. As the philosopher-activist Lisa Kemmerer puts it (2012: 115; see also Harris 2006: 209), from a Buddhist perspective we are 'just one birth away from the life of a mountain bluebird or desert toad'.[5] The Buddha was even closer; as the opening and the closing paragraphs of the tale state, the Lord identified 'himself with every living creature' (Khoroche 2006: 6, 9).

We need not to be Buddhist 'true believers' to accept their, and the jātakas', point that humans and other animals are interdependent or co-dependent, which is drilled into the tale's listeners or readers rather dramatically by the Bodhisatta's act of self-sacrifice. Taking this as a social act, as I have suggested above, the tale can be interpreted as an example of sociologist Marcel Mauss's (1969) way of understanding gift-giving, where the point is to build solidarity and to strengthen mutual relationality. Mauss famously concentrated on human–human relations, but the tale expands solidarity beyond mere humans (cf. Ohnuma 2005: 105–7).

At this point, we must return to Ajita wondering about the Bodhisatta's 'extraordinary deed'. Despite my earlier claims to the contrary, a sceptic, or even a Buddhist, might argue that feeding oneself to a hungry tigress is beyond common human ability. This objection was anticipated in the jātaka collection Mahāvastu (Great Story), which consists of various texts on the Buddha's life histories, and dates from around the first centuries BCE to the first centuries CE. In the Mahāvastu version, the Mahāsattva, i.e., the Buddha, says that self-sacrifice is 'difficult for people like us, who are so fond of our lives and bodies, and who have so little intelligence. It is not at all difficult, however, for others, who are true men, intent on benefiting their fellow-creatures [...]. Holy men are born of pity and compassion' (Conze 1984: 21).

Supposing that one is not that holy, what *can* people do to show compassion? At least we may accept that in an interdependent world we are all socially as well as morally responsible for each other (cf. Birke and Hockenhull 2012: 20–3). The tale stresses that such responsibility rises in relation to the other and has to be unconditional or compassionate, as the Buddhists put it, and applicable to species other than human species. But how can this be possible? And why should one bother to treat the other in a compassionate and relational manner?

The Bodhisatta said that if he could help yet refused to do so, he would be committing a crime (Khoroche 2006: 7–8). A Buddhist might call his attitude and action an expression of *adrstaphala*, relating to the other in a manner that leads to an 'unreciprocated' gift that 'bears fruit in the transcendent future' (Ohnuma 2005: 106). Put another way, by doing what he did, the Buddha-to-be merely followed the normal, natural course of life. In the modern Western tradition, philosopher Emmanuel Levinas (1961) emphasized the built-in asymmetry in human relations by pointing out that they are vulnerable, situational, and constantly changing. The point is that ethics, as well as human–animal relationality in general, is not about beings as such (humans or otherwise) but about relations.[6] It is of less importance whether the Bodhisatta, or you, or I, or anyone else, is generous as an individual. What matters is striving for creating compassionate relations.

In my view, Reiko Ohnuma has something like this in her mind when she notes (2005: 110) that there is no truly unreciprocated gift or action; whatever our conscious intentions, our relations to others always involve some desires and expectations. In other words, 'I' do not exist alone, but in relation the others, and emotions are perhaps the best indicators of that relationality. The 'Tale of the Tigress' acknowledges this by stating that the Bodhisatta's compassion for distressed creatures made him shake like the Himalayas in an earthquake (Khoroche 2006: 7). With such emotional commitment he showed himself ready to invest himself in relationality, and indicated that his involvement was not selfish but rather aimed at what may be termed a 'building of common benefit' or, to put it more simply, companionability.

Relating companionably to the tigress

What does companionability mean? To sketch it in Buddhist terms, one may speak of compassion, or *karuna* (Keown 2004: 138). In my parlance, the term means that one both keeps company with someone and takes her or him as an equal partner, despite the fact that the relationship is not, and cannot be, fully equal. Companionability also includes the notion that one is not only able to imagine oneself in another's position, but also to a significant degree 'knows' the other's mind, i.e., understands her ways of thinking and feeling.

How is this possible? Cognitive psychologists tell us that one can to some extent 'read' the other's mind because we have an inborn capacity to 'attribute intentions to others and respond empathetically' (Duffy 2011: 205). This capacity is called the 'theory of mind'. Without using the term, the autist ethologist Temple Grandin applies the idea when she argues that she is able to understand animals (in her case cows), because she and the cows both think in pictures. She has substantiated her claim by designing (from a human viewpoint) animal-handling equipment (Grandin and Johnson 2005). The jātaka uses a reminiscent theory of mind. The tale implies that the Bodhisatta takes the position of both the tigress (by understanding that she is hungry) and the cubs (assuming that they are terrified), and can therefore relate to both of them in a compassionate, or companionable, manner.

Read in this way, the 'Tale of the Tigress' suggests that companionability is something social (it is about relations), existential (it is about the meaning of one's behaviour), and ethical (it is about generosity). To apply the thinking of Ian Harris (2006: 213), in the tale the Bodhisatta understands that all sentient beings are caught in the endless circle of *saṃsāra*, the cycle of births and rebirths, which makes all beings a related community struggling for liberation. In other words, all beings live in comparable existential conditions. Having established that, Harris (ibid.) continues by claiming that generosity or human loving-kindness towards other animals 'can happily be taken along as baggage on the path to perfection', but at some stage a true Buddhist has to leave it 'at the side of the road', because a liberated being no longer needs such crutches. Therefore, for Harris, generosity is a means, not a characteristic. Mahāyāna schools call this view *upāya-kauśalya*, skilful means (Keown 2004: 318). This means relating to the other with the Buddha's ingenuity in order to alleviate the other's suffering.

I agree that, insofar as the notion of 'skilful means' indicates that the Buddha's teaching has to be rethought, Harris's view is a correct one. Yet, I disagree with the concept that generosity to other animals is something that could or should be disregarded at the threshold of liberation. While Harris's view accords with a traditional understanding of Buddhism, there are other, less trodden paths which I consider worth tracking. The tacit question, of course, is whether we can do so without distorting the Buddhist tradition. I believe that this may be possible, because what is called the Buddhist tradition is, in fact, not a monolith; as I have stated above, for example, notions on the Buddha can be based on three 'lineages' and their various combinations. Thus it should be possible to reconsider Buddhist views without abandoning the Buddhist dharma. This has also been done. Regarding jātakas, Ivette Vargas points out (2006: 219) that in the Indian and Tibetan versions animals, such as the above-mentioned quail, are portrayed as active agents or persons, not as passive objects, which justifies a companionable relationship with them.

Ignoring the fact that jātakas are not considered first class texts of Buddhist dharma, and keeping in mind that they nevertheless represent a mainstream Buddhist lay tradition, I think that Vargas is correct when she suggests that rather than trying to explain away inconsistencies in different traditions of human–animal relations, we should instead take them as potential options. In other words, she, too, suggests the use of skilful means, although her view differs from that of Harris.

But why are her, or my, skilful means best? In my view, the main difference concerns relationality. While Harris suggests concentrating chiefly, or exclusively, on one's own liberation, Vargas prefers relating to the other with the Buddha's ingenuity in alleviating the other's suffering. As to the 'Tale of the Tigress', this means that while at the time of the Buddha the emphasis on the human part of the relation was appropriate, today, due to the changed or deteriorated conditions of other animals (in factory farms, for instance), it is no longer so. An increase in suffering requires an increase in compassionate relationality.

Some scholars may find my stress on human responsibility in relation to other animals unconvincing. Do I merely anthropomorphize (other) animals by claiming that they, too, are moral subjects (cf. Hauser 2006: 505)? Is companionability but an unnecessary human intervention in the order of nature? For example, animals in nature do eat each other, including their own progeny. By preventing this, the Bodhisatta thus 'broke' the laws of nature. Other scholars might claim that any debate of animals' morality is futile, because only humans have, during their evolutionary history, invented moral reasoning and rules that have allowed them 'to survive in a world where the range of temptations evolves at an exponential pace' (Hauser 2006: 515; cf. Rowlands 2012: 3–7, 18–22).

I disagree with those who argue that morality does not matter in human–animal relations. I side with Levinas (1961) that relationality always brings with it a sort of responsibility, and argue that relations presume reversibility. The 'Tale of the Tigress' and similar stories can be read as if the moral reasoning and rules we allegedly invented are part of other animals' perspectives, too. In practice this means that humans accept other animals as potential moral agents and treat them accordingly. It does *not* mean that one does, or should, anthropomorphize animals. The Bodhisatta evidently did not consider the tigress as a fellow human. Quite the contrary: he preserved her 'animality', or accepted her total difference, by acknowledging her 'right' to eat meat. Thus he took her as seriously as his fellow-being, with compassion, or empathy (Rowlands 2012: 32–8). A Buddhist would say that the Bodhisatta thus produced meritorious karma (cf. Strong 2011: 179). I speak of companionability in order to stress not the *result*, as a Buddhist might do, but the *process*, the relational interaction of human with non–human animals.

From a Western, rights-based view, Buddhist compassion, and my companionability, may seem dubious. Sociologist Erik Cohen (2013: 278) correctly notes that 'though Buddhism teaches a compassionate attitude towards animals, it does not recognize that animals have autonomous rights'. The reason, according to Cohen, is the built-in hierarchy in the steps leading to enlightenment: animals stand lower on the steps than humans do. And he indeed has a point. In Dakkhināvibhanga Sutta,[7] 'the Buddha tells his attendant Ānanda that giving [being generous] to an animal produces ten times less merit than giving to an evil human being' (Ohnuma 2005: 114). While I agree that this is a problem, I also think that Cohen makes unwarranted generalizations based on his single example, the modern Thai Tiger Temple (actually a zoo); the animals suffering there do not invalidate all Buddhist philosophy and ethics. As Cohen (2013: 281) also writes, 'Buddhism recognizes that animals suffer from human cruelty'. As I have suggested above, this can be read as one version of the theory of mind: humans have a shared (biological and existential) basis with other animals, and emotions and reason tell us when others are suffering. Companionability can be based on such understanding, on a belief that relations that make life less painful are generous to everyone. Thus companionability may be summed up as building relations that aim at treating the other as a moral agent, and being ready (to the point of offering one's life) to alleviate the other's suffering. These actions appear general enough to be expanded from the context of the 'Tale of the Tigress', and the Buddhist ethics in general, to inform the creation and maintenance of other human–animal relations and, perhaps, even trans-species identities (cf. Deleanu 2000: 123–4; Robertson 2013: 13).

Notes

1 It is one of the three canonical Theravāda scriptures. The other two are Vinaya Pitaka, or rules for monks and nuns, and Abhidharma Pitaka, commentaries of *sūtras* and other philosophical material.
2 According to strict Theravāda tradition, animals are incapable of enlightenment.
3 I prefer the Pāli term to refer to the historical Buddha in his previous births. The Sanskrit term bodhisattva also embraces other beings.
4 A sūtra (in Pāli, a *sutta*) is a discourse of the Buddha.
5 The notion is based on Samyutta Nikāya (Connected Discourses), which is a part of the Sutta Pitaka, and some Mahayana texts such as the Lankāvatāra Sūtra (see Chapple 1997: 143), which emphasizes consciousness as the only true reality and figures prominently in Far Eastern Buddhism.
6 The question of morals in (other) animals is of course highly disputed. See Rowlands (2012: 9).
7 One of the 152 'middle length discourses' making up Majjhima Nikāya, a part of the Sutta Pitaka.

References

Appleton, N., 2010. *Jātaka stories in Theravāda Buddhism: narrating the bodhisatta path.* Farnham, UK: Ashgate.

Birke, L. and Hockenhull, J., 2012. On investigating human–animal bonds: realities, relatings, research. *In:* L. Birke and J. Hockenhull, eds. *Crossing boundaries: investigating human–animal relationships.* Leiden, Netherlands: Brill, 15–36.

Chapple, C.K., 1997. Animals and environment in the Buddhist birth stories. *In:* M.A. Tucker and D.R. Williams, eds. *Buddhism and ecology: the interconnection of dharma and deeds.* Cambridge, US: Harvard University Center for the Study of World Religions, 131–48.

Cohen, E., 2013. 'Buddhist compassion' and 'animal abuse' in Thailand's Tiger Temple. *Society and Animals,* 21 (3), 266–83.

Conze, E. (trans.), 1984 [1959]. *Buddhist scriptures.* Harmondsworth, UK: Penguin Books.

Deleanu, F., 2000. Buddhist 'ethology' in the Pāli Canon: between symbol and observation. *Eastern Buddhist,* 32 (2), 79–127.

Duffy, J., 2011. The pathos of 'mindblindness': autism, science, and sadness in 'theory of mind' narratives. *Journal of Literary and Cultural Disability Studies,* 5 (2), 201–16.

Gombrich, R., 1977. Introduction. *In:* M. Cone and R. Gombrich, eds. *The perfect generosity of Prince Vessantara: a Buddhist epic.* Oxford, UK: Clarendon Press, xv–xlvii.

Grandin, T. and Johnson, C., 2005. *Animals in translation: using the mysteries of autism to decode animal behavior.* New York: Scribner.

Green, S., 2006. *Tiger.* London: Reaktion Books.

Harris, I., 2006. 'A vast unsupervised recycling plant': animals and the Buddhist cosmos. *In:* P. Waldau and K. Patton, eds. *A communion of subjects: animals in religion, science, and ethics.* New York: Columbia University Press, 207–17.

Hauser, M., 2006. Are animals moral agents? Evolutionary building blocks of morality. *In:* P. Waldau and K. Patton, eds. *A communion of subjects: animals in religion, science, and ethics.* New York: Columbia University Press, 505–18.

Kemmerer, L., 2012. *Animals and world religions.* Oxford, UK: Oxford University Press.

Keown, D., ed., 2004. *Oxford dictionary of Buddhism.* Oxford, UK: Oxford University Press.

Khoroche, P., (trans.), 2006 [1989]. *Once the Buddha was a monkey: Ārya Śūra's Jātakamālā.* Chicago, US: The University of Chicago Press.

Levinas, E.E., 1961. *Totalité et infini: essai sur l'extériorité*. The Hague: Martinus Nijhoff.

Matsumura, J., 2010. The Vyāghrī-Jātaka known to Sri Lankan Buddhists and its relation to the Northern Buddhist versions. *Journal of Indian and Buddhist Studies*, 58 (3), 1164–72.

Mauss, M., 1969. *The gift: forms and functions of exchange in archaic societies*. Trans. Ian Cunnison. London: Cohen and West.

Ohnuma, R., 1998. The gift of the body and the gift of dharma. *History of Religions*, 37 (4), 323–59.

Ohnuma, R. 2005. Gift. *In:* D.S. Lopez, ed. *Critical terms for the study of Buddhism*. Chicago, US: The University of Chicago Press, 103–23.

Robertson, V.L.D., 2013. The beast within: anthrozoomorphic identity and alternative spirituality in the online therianthropy movement. *Nova Religio: The Journal of Alternative and Emergent Religions*, 16 (3), 7–30.

Rowlands, M., 2012. *Can animals be moral?* New York: Oxford University Press.

Sheravanichkul, A., 2008. Self-sacrifice of the bodhisatta in the Paññāsa Jātaka. *Religion Compass*, 2 (5), 769–87.

Strong, J., 2011. The Buddha as ender and transformer of lineages. *Religions of South Asia*, 5 (1–2), 171–88.

Van Deusen, K., ed., 2001. *The flying tiger: women shamans and storytellers of the Amur*. Montreal, Canada: McGill-Queen's University Press.

Vargas, I., 2006. Snake-kings, boars' heads, deer parks, monkey talk: animals as transmitters and transformers in Indian and Tibetan Buddhist narratives. *In:* P. Waldau and K. Patton, eds. *A communion of subjects: animals in religion, science, and ethics*. New York: Columbia University Press, 218–40.

12 Passing the cattle car

Anthropomorphism, animal suffering, and James Agee's 'A Mother's Tale'

Jouni Teittinen

Anthropomorphism, or the projection of human attributes on the non-human realm, is a concept that often surfaces in discussions of both fictive and non-fictive representations of non-human animals. However, representations of animals and animal experience should not only be taken as descriptive. As I will argue in connection with the American writer James Agee's short story 'A Mother's Tale' (1968 [1952]), they can also be performative in ways that have consequences for how the limits of anthropomorphic practice are negotiated. Instead of attempting an in-depth interpretation of Agee's multi-layered text as such, I will concentrate on articulating some of the problems and possibilities involved in empathetically engaging with such animal stories.

Taking my cue from Robert McKay's (forthcoming) reading of Agee's short story, I first acknowledge the complications the text presents for the conventionally allegorical and the trivially anthropomorphic readings. I will then elaborate the ambiguities inherent in the concept of anthropomorphism, and argue that anthropomorphism is not only or even primarily an epistemic issue but also a matter of deeply affective ingroup–outgroup control. I suggest that this problem of 'us' versus 'them' is closely connected to the recognition of suffering, and that representations of pain in particular have a special place, however problematic, in readjusting our sense of commonality and communality with non-human animals. The renewal of these felt relations does not do away with problems of representing non-human life, but it may markedly shift their often taken-for-granted coordinates.

Perspectives on 'A Mother's Tale': varieties of anthropomorphism

'A Mother's Tale' is the story of a mother cow (or a dam), who is confronted with the task of explaining to young calves why the cattle from the next pasture is being herded away. The central part of the narrative consists of the tale the mother tells, a folk legend of sorts that has gained mythical proportions as it has been passed on by numerous bovine generations. The tale documents a single steer's (eventually mythologized as 'The One Who Came Back') long journey in the cattle car to the massive stockyards, his confrontation with the 'Man With the Hammer', and his eventual escape back to his home farm to reveal the truth about the train, the humans, and the grim fate of earthly (bovine) existence.

Quite obviously, the story is anthropomorphic, beginning with the title which plays with the readers' expectations by not differentiating between a mother cow and a human mother. To name a story anthropomorphic does not by itself mean very much, however, since anthropomorphism is an unusually slippery term. As it turns out, 'A Mother's Tale' is in fact a veritable study in anthropomorphism; not only for the fact that the narrative is thoroughly anthropomorphic, but for the varieties of anthropomorphism it makes use of, and for the complications it casts upon the notion of anthropomorphism itself.

In the first instance, 'A Mother's Tale' has elements of what John Simons (2002: 119) calls 'trivial anthropomorphism': the kind of writing which, without carrying a fable-like moral, 'treat[s] animals as if they were people' but 'does not press against and force us to question the reality, or otherwise, of the boundary of the human and the non-human'. This reading is suggested by the many transparently human conceptions the animal characters harbour, as when the mother explains the rationale for taking her story to its gruesome conclusion: 'She always tried hard to be a reasonably modern mother. It was probably better, she felt, to go on, than to leave them all full of imaginings and mystification' (Agee 1968: 224).

More interestingly, 'A Mother's Tale' reads as an anthropomorphic allegory. As Robert McKay (forthcoming)[1] notes, the short story has variously been read

> as an allegorical fable dealing with human destiny, as symbolic fable about Agee's own mortality, a polemic against war, a skeptical commentary on the individual struggle with modern society, Agee's most refined statement on the nature of art and the nature of truth, and the moral lesson that it's proper to rail and struggle against fate, but that fate can't be avoided.

However, as McKay (forthcoming) has also shown, there is more to 'A Mother's Tale' than trivial anthropomorphism and allegory. McKay finds two key factors in the story that complicate the allegorical reading: the intensive, detailed descriptions of the animals' bodily experience, and the dramatic irony that forces the reader to confirm those truths about animal production that remain only a suspect legend for the tale's animal audience (ibid.). If allegory is considered along the lines of Simons's (2002: 119) general definition of a fable, in which 'there is no stage at which a reader can doubt, or is invited to doubt, that what he or she is being offered is a tale which explores the human condition' it becomes clear that 'A Mother's Tale' does not qualify as a strict allegory. The animal-centred reading does not imply that we ought to abandon the allegorical dimension; on the contrary, Agee's short story gets a crucial momentum out of this tension.

Here it is necessary to note that there are two distinct ways of understanding the notion of anthropomorphism. In its general and casual sense, the notion can simply refer to any instance in which an aspect of the non-human is constructed to appear as (supposedly) human, no matter the intention or context. In a more specific sense, the notion is used to point out instances of what we might call 'anthropomorphic thinking'. These practices of thought either unintentionally blend human and animal characteristics or intentionally challenge their boundaries, making anthropomorphism not only a textual or a heuristic device (as in trivial anthropomorphism or allegory) but a more complex issue for representation and thought.

A strictly allegorical presentation can be anthropomorphic in the more general sense, but as such it does not involve anthropomorphic thinking. This is because a purely allegorical animal figure is not even supposed to convey animal life (although these animal allegories might create some ground for, or come out of, a corresponding anthropomorphism in thought): to read the anthropomorphic display of 'A Mother's Tale' as purely allegorical is to completely bracket the animals themselves. Anthropomorphic thinking, in turn, is opposed to allegory, since it needs the animals *as* animals in order to wittingly or unwittingly confuse the human–animal divide. Thus for epistemic as well as ethical questions, it is specifically anthropomorphic thinking that carries a special interest.

Commonly, the very concept of anthropomorphism (as sustained anthropomorphic thinking) is thought to implicate an error, a false attribution of human characteristics on non-human entities (cf. Sober 2005: 85; Tyler 2009: 15) – a variant of the 'pathetic fallacy' of projecting feelings where there are none (Garrard 2004: 36). But anthropomorphism may also be accepted as an occasionally valid and sometimes inevitable practice (cf. Sober 2005: 85; Pollock and Rainwater 2005: 14). When the practice of projecting human attributes onto non-humans is not by definition seen as faulty but merely as problematic, it may begin to serve a destabilizing, disconcerting function. This is the prospect raised by 'A Mother's Tale'. In addition to its elements of both trivial anthropomorphism and anthropomorphic allegory, 'A Mother's Tale' presents a case – again in Simons's (2002: 120) terms – of 'strong anthropomorphism', which employs anthropomorphism 'to show how the non-human experience differs from the human or to create profound questions in the reader's mind as to the extent to which humans and non-humans are really different'. It is this sense of anthropomorphism that most concerns me here.

From epistemic doubt to shared vulnerability

The late nineteenth-century principle known as Morgan's canon states that we must assume only those minimal mental characteristics that are absolutely necessary for interpreting an animal's behaviour; ideally, the explanation should not include reference to any mental or psychological factors (Sober 2005: 86). This contested but still standard critique of anthropomorphism need not deny all similarity between humans and non-human animals, since the criticism finally rests on an epistemic basis. As our knowledge about the mental lives of non-human animals eventually rests on speculation and uncertain analogy, the critics say, we should keep silent.

The heart of the issue, then, is not that animals are not at all like humans, but rather the problematic nature of this likeness. Granted that it is there, how do we measure, map and delimit this territory? Since we have no 'access' to other animals' experience, we easily find descriptions of such experience suspect. To quote, once again, John Simons (2002: 116): 'To portray nonhumans as if they were humans is to bring them into a discursive realm in which it is possible to give the illusion that their experience is being reproduced', while in fact we are always dealing with construed representations.

Consider the way 'A Mother's Tale' describes the cattle car's unfamiliar movement as the steer protagonist is being transported from the farm:

> He felt as if everything in him was *falling*, as if he had been filled full of a heavy liquid that all wanted to flow one way, and all the others were leaning as he was leaning, away from this queer heaviness that was trying to pull them over, and then just as suddenly this leaning heaviness was gone and they nearly fell again before they could stop leaning against it (Agee 1968: 228; emphasis original).

Then, as the cars finally arrive at the stockyard and the livestock is let out:

> he said that it turned from night to day and from day to night and back again several times over, with the train moving nearly all of this time, and that when it finally stopped, early one morning, they were all so tired and so discouraged that they hardly even noticed any longer, let alone felt any hope that anything, would change for them, ever again; and then all of a sudden men came up and put up a wide walk and unbarred the door and slid it open, and it was the most wonderful and happy moment of his life when he saw the door open, and walked into the open air with all his joints trembling [...] (Agee 1968: 230).

The descriptions might be seen as purporting to diminish the sense of distance between representation and experience (and so to 'reproduce' animal experience) by evoking vivid bodily, physical, and spatial movements in accordance with the supposedly simple elements of animal experience. This is accompanied on the structural level by 'childlike' repetitiveness in rhythm and wording, in the latter example especially the paratactic linking of sentences ('and... and...') to avoid complex syntax and the repetition of the word 'open' to convey the overwhelming experience of exiting the car.

Should these scenes be considered as ways of portraying animals, in Simons's (2002: 116) turn of phrase, 'as if they were humans'? In line with Simons's problematic opposition between representation and reproduction, the philosopher Thomas Nagel (1974: 439) has famously decreed that it is impossible for us to really imagine the experiential worlds of animals *as those animals experience it*, since we always end up thinking what it would be like *for us* to be that animal. To take the cattle car scenes as presenting actual animal experience might be an example of just such a fallacy: this is what I would go through if I was a cow in that car. In the passages quoted above, the fallacy is paralleled by the fact that, however skilfully the steer's point of view is focalized, we cannot rid ourselves of the irony involved in knowing where and why the cattle are travelling.

Accordingly, if we grant (along with Nagel and good reason) that with many animal species it surely is 'like something' to be them, it seems we must simultaneously assume some commonality with human experience and refrain from saying anything much of it. Faced with this impasse, one option is simply to change the topic of conversation and pay little heed to problems of projection, representation, and criteria for knowledge. Instead, accounts of animal life like the one presented in 'A Mother's Tale' can be taken as ways to move, to affect, and so to make the reader experience through the perhaps anthropomorphizing accounts the quite factual distress and vulnerability of the animals (for a sympathetic overview of the discussion, see Thierman 2011). The important thing from this perspective is not so much to think of animals as like us, but to come to realize our own, shared, very animal condition.

What is commendable in the 'shared vulnerability' approach is its willingness to face the actual, messy encounters between humans, animals, and their mediating (in this case textual) environments. Nevertheless, perhaps we should not surrender the epistemic claim to the sceptic and rest content with evoking a more fundamental vulnerability. Instead, closer attention should be paid to the ways in which epistemic and ethical questions necessarily intermingle. This is not necessarily a matter of placing ethics (taken as a theory of norms) foremost and free of epistemic worries; rather, we ought to acknowledge that these worries are in part conditioned by ethics, especially when ethics is understood in its larger sense of attitudes, habits, and the spirit of a culture (the culture's *ethos*). In other words, we should not strictly separate the affective element from claims to knowledge, since new affective connections make possible the formation of new attitudes towards what makes sense to us in the first place, and what sort of claims we consider worthy of interrogation.

In the community of the suffering

In its critical usage the notion of anthropomorphism regulates what is proper to man and not to animal. In this sense, it is a conceptual tool that operates by sketching and guarding the borders of the human. But how is this 'human' delimited? While we might in principle make infinite careful demarcations, in its actual critical employment the notion of anthropomorphism gains its force from the implicit assumption that the categories of man and animal, like their respective relevant properties, are relatively apparent to begin with and need only to be applied to the matter at hand. In other words, in its simple form the accusation of anthropomorphism has to assume a simple dividing line in order to function as such. Herein lies the concept's inner circularity and instability, its inability to insulate itself in the realm of knowledge claims. We might have a biological understanding of the basic differences between man and other animals, but a critically operational notion of anthropomorphism also requires a culturally constructed understanding of what it means to be human, as opposed to being an animal. As Cora Diamond (1995: 330–1) puts it, 'what is meant by doing something to an animal, what is meant by something's being an animal, [...] [is] not given for our thought independently of such a mass of ways of thinking about and responding to them'.

The human is both a biological given and a politically loaded concept, and here things get easily muddled. As is often noted, the human is a constitutively exclusionary category not only between species but also within biological humanity. Max Pensky (2008: xii), for example, claims that '[c]osmopolitan solidarity [...] cannot coherently base its normative claims on the notion of an inclusion without exclusion. Even the concept of "the human" presupposes the ongoing work of patrolling, revising, contesting, and enforcing exclusionary boundaries'. Animalization, of course, is a standard figure in racist and chauvinist rhetoric. At the same time, many non-human animals are granted a place in our midst as an index of their 'human' characters; above all our pets, but one also thinks for example of the 'limited human rights' granted to great apes by the Spanish parliament in 2008 and to a single orang-utan named Sandra by an Argentine court in 2014.

While it is important not to indiscriminately confuse the biological human–animal distinction with the human as a structure of political inclusion and exclusion, in the matter of anthropomorphism these themes may substantially overlap. As the biological notion of species cannot by itself decide the critical issue of whether supposedly human properties are falsely ascribed to non-human animals, the required additional and implicit demarcation seems to be one of 'us' versus 'them'. The more we regard animals – of this or that sort, or perhaps only this or that individual – as 'one of us', the more comfortable we are with accepting such descriptions of them that we otherwise would deem anthropomorphic.

The experience of 'us-ness', of community and communality, appears to have an intimate and perhaps essential connection with acknowledging the suffering of others: who or what can appear as someone or something whose suffering we are able to recognize as meaningful and as something that, perhaps despite a certain ineliminable alterity, places a demand on us. One could here speak of the community of the suffering, of those whose pain – or fear, or death – brings us to grief.

How, then, should the constitution of this community be understood? An intuitive view is that a proper recognition of suffering has to be preceded by a more originary identification. Matthew Potolsky (2002: 122), for example, paraphrases Freud as noting that 'feeling is a result of identification, not its cause [...]. We sympathize with those who correspond to our identifications, rather than identifying with those whom we pity'. But there is another view available, which does not see the recognition of suffering as simply resulting from already belonging to this or that community, whose contours could be defined independently of the question of suffering. As Judith Butler has argued, we can see the sympathetic experience itself working upon and affecting the limits of that community. In the words of James Stanescu (2012: 578), who applies Butler's thinking to the question of animals, 'It is our very ability to be wounded, our very dependency, that brings us together' – in other words, it is our shared vulnerability preceding sociality and social identification, that forms what Stanescu (2012: 575) and Butler (2009: 19–20) consider a social ontology that grounds community (for a systematic argument on how responding to suffering grounds human community, see Taylor 2002).

However, if we examine how this imagined community might establish more inclusive human–animal relations, vulnerability and suffering have problems as baseline concepts. Vulnerability has a slightly abstract feel – it is no doubt our actual condition but also a potentiality, a capacity or incapacity of sorts. Suffering as its constitutively necessary co-concept is clearly more concrete, but retains an essential ambiguity about the nature of the suffering: it entails loss, but not necessarily very much more, and like vulnerability (or precariousness, Butler's term of choice) can be seen as something worth embracing up to a point, especially when employed abstractly. While vulnerability and suffering cannot properly be disassociated from being embodied and mortal, they can also be conceived of as non-bodily states. Consequently, it is possible to maintain that while animals feel bodily pain, they do not experience suffering or vulnerability proper, since mentally they lack something or other to, as it were, inject pain with deeper meaning.

It is for these reasons at least that physical, bodily pain can gain a fundamental status when considering the community of the suffering. Pain, precisely, matters without the need for being accompanied by anything extra. This finds support in the way Katerina Kolozova (2010) emphasizes the bodily ground of Butlerian vulnerability. According to Kolozova (2010: 16), the 'revolutionary potential' in Butler's claim that we can identify with 'suffering itself' lies in the way it 'enables inclusion unlimited by the inclusiveness of the category of Human'. As Kolozova (2010: 118) suggests:

> Bodily suffering, a body in utter helplessness facing a threat of brutal violence, is an instance we can identify with without any need of conceptual frame that would enable valorization or 'making sense' of it, i.e., without the category of humanity.

If it is the case that being confronted with bodily pain carries an element that at some level short-circuits whatever discursive demands we might pose to its intelligibility, we can understand the significance of this confrontation for identifying with the animal's experience. (In addition to the felt effect of confronting an animal in pain, it is not irrelevant that there is both scientific and popular consensus that many animals – with birds and mammals at the centre – not only exhibit pain behaviour but also experience pain.)

Of course, we cannot actually share the animal's physical pain, but equally we cannot strictly speaking share the pain of another human (for this reason I find Kolozova's [2010: 144] concept of 'co-suffering' somewhat suspicious). As Elaine Scarry (1985: 3) puts it in her seminal study *Body in Pain*:

> Whatever pain achieves, it achieves in part through its unsharability, and it ensures this unsharability through its resistance to language. [...] Physical pain does not simply resist language but actively destroys it, bringing about an immediate reversion to a state anterior to language, to the sounds and cries a human being makes before language is learned.

Even though Scarry seems too strictly to equate communication with written or spoken language, her insight into the fundamental unshareability of human pain can gesture us towards thinking of animal pain in connection with the problem of anthropomorphism. That is, the fact that pain (perhaps more than some other modes of suffering) defies representation can make us realize that our relation to another's pain as well as to suffering more generally should not be thought of in the first instance as epistemic. We can here draw inspiration from Cora Diamond's (1995: 339) observation:

> it is not a fact that a titmouse has a life; if one speaks that way it expresses a particular relation within a broadly specifiable range to titmice. It is no more biological than it would be a biological point should you call another person a 'traveler between life and death': that is not a biological point dressed up in poetical language.

Similarly, to say that someone or something is in pain is not to remark on biological activity in that individual's C-fibres (outside perhaps a clinical context), or to make a knowledge claim in general, but to take on a particular relation to her or his or its experience. However inadequate and consciously or unconsciously played down this relation might be, it is usually essentially one of identification or sympathy.

Consider the following graphic description given in 'A Mother's Tale' of what the steer suffers through as the result of a failed killing:

> How long he lay in this darkness he couldn't know, but when he began to come out of it, all he knew was the most unspeakable dreadful pain. He was upside down and very slowly swinging and turning, for he was hanging by the tendons of his heels from great frightful hooks, and he has told us that the feeling was as if his hide were being torn from him inch by inch, in one piece. And then as he became more clearly aware he found that this was exactly what was happening. Knives would sliver and slice along both flanks, between the hide and the living flesh; then there was a moment of most precious relief; then red hands seized his hide and there was a jerking of the hide and a tearing of tissue which it was almost as terrible to hear as to feel turning his whole body and the poor head at the bottom of it; and then the knives again (Agee 1968: 235).

While some details of the passage may come across as anthropomorphic more readily than others, and there is the mythical and even biblical injunction 'he has told us' to remind the reader (as if for a moment to pause the pictorial onslaught) that the narrative is doubly mediated, this is probably not where the reader's attention is. The image of pain in itself is not easy to dismiss as anthropomorphic; it leaves little room for words.

The trouble with pain

There is something undeniably problematic in the focus on animal suffering and specifically pain. As Donna Haraway (2008: 19–22) has argued, emphasizing animal suffering tends to mistakenly strengthen the conventional understanding of the fundamental passivity and lack of agency of the animals (in contrast to the active human). Emphasizing bodily pain has the added risk of making the animals themselves appear as nothing but bodily, with the false corollary that it is enough for the animals' well-being to ensure them a physically painless environment. Robert McKay (forthcoming) notes that to keep our attention primarily on bodily suffering as the 'sphere that commands ethico-political recognition' (as has been the standard in animal welfare policy) is to turn it away from other, perhaps more fecund forms of ethical engagement. McKay's central and well-based claim, in fact, is that 'A Mother's Tale' helps us imagine ways to go against a suffering-centred conception of animal experience (ibid.).

While we surely have ways to relate with animals that do not depend on pain and suffering, it seems to be the case that it is too early to abandon representations of pain as a central means to empathize with much of non-human experience. To witness a suffering animal is not enough, but if we consider the complex triangle of anthropomorphism, communality, and suffering, the encounter with pain, so to speak, gets a foot in the door. From affect and sympathy we return (and often in haste) to our more composed, casual, and articulate relation to the suffering of animals, reclaim the distance that allows us to consider the matter of anthropomorphism and its reasonable limits. But these limits, forged through a feeling of communality, may have shifted, and with them our attitude towards various aspects of human–animal relations – including those that allow animals a more active role in our midst.

Another cautionary note is in order. While emphasizing the role of bodily pain and vulnerability in communal identification, it is crucial to notice the ways in which this vulnerability itself becomes mediated in society. First, it is difficult to separate the immediate-seeming physical recognition of the body in pain from the way our perceptions are culturally conditioned; second, as Kolozova (2010: 150–2) notes, this recognition has to always enter into the realm of discourse to evolve into actual solidarity and political action.

After all, it is a commonplace that we are oblivious even to human suffering when it is sufficiently removed – how fast we pass by the atrocities we are fed with on the news – since grief is anyway on its way and one has to live, has not one? So too our ability to confront animal pain has to do with the limits and habits of cultural intelligibility, and with the technologies of mediation guiding those habits. As Judith Butler (2004: xx–xxi) writes in the context of the war in Iraq:

> The public sphere is constituted in part by what can appear, and the regulation of the sphere of appearance is one way to establish what will count as reality, and what will not. It is also a way of establishing whose lives can be marked as lives, and whose deaths will count as deaths. [And, I will add, whose suffering will count.] Our capacity to feel and to apprehend hangs in the balance. But so, too, does the fate of the reality of certain lives and deaths as well as the ability to think critically and publicly about the effects of war.

From the perspective of this chapter, it is important to recognize that what is at stake here is not only the ability to think critically about animal life and death, but also our very 'capacity to feel and to apprehend' (Butler 2004: xxi), and that these vectors of affective and critical engagement cannot be in their entirety disentangled.

In this vein, there is something very appropriate in the following scene from 'A Mother's Tale', again describing the steer's journey in the cattle car.

> They were standing still, and cars of a very different kind began to move slowly past. These cars were not red, but black, with many glass windows like those in a house; and he says they were as full of human beings as the car he was in was full of our kind. And one of these people looked into his eyes and smiled, as if he liked him, or as if he knew only too well how hard the journey was.
>
> 'So by his account it happens to them, too,' she said, with a certain pleased vindictiveness. 'Only they were sitting down at their ease, not standing. And the one who smiled was eating.' (Agee 1968: 229)

The passage serves as a good example of how the allegorical and the literal complement complicate each other in Agee's text. At first reading, the scene appears to serve an allegorical interpretation, and even the mother's (who narrates the scene) somewhat sardonic reference to the passenger's ease and dining (which, we might surmise, is just what indirectly puts the cattle on the train) can be taken to imply that all their comfort cannot protect humans from the eventual hardness of their journey.

Significantly, this underemphasized key scene is also the closest thing to a real human–animal connection that the story allows. Indeed, this is the only occasion where an individual is singled out from the faceless multitude of men handling the livestock, save for the apocalyptic and inhuman figure of the 'Man With the Hammer' at the slaughterhouse. The reader is given the chance to briefly identify with a person looking at and perhaps empathizing with the animals.

While the steer takes the passenger's smile as indicating sympathy, however, it is not hard to envision other options. The emphasis on the fact that he is eating again evokes a dramatic irony. Perhaps the passenger does empathize, but gives no thought to the (assumed) connection between the cattle car and his plate; perhaps he just enjoys his steak. Perhaps he smiles precisely because he understands this connection *and* likes his steak. It is of course possible that the man smiles, as the steer imagines he might, because he thinks he understands the hardness of the cattle's journey, and even feels that he in some way shares the livestock's creaturely and mortal ride. The reference to the passenger's ease, however, casts an ironic light on the latter option.

In all of these readings (notwithstanding the plainly allegorical), the passage can be seen to reflect the problem of attempting to approach animal experience – say, through a literary text – and the related claim to knowledge and understanding. On both sides one can find an attempt to understand and empathize, but the steer and the man remain in separate spaces, any identification mediated by more than just glass windows.

In conclusion

'A Mother's Tale' is not the sort of anthropomorphic animal story where men and animals converse with each other with little difficulty. The humans in the story are inexplicable to the animals, their purposes and thoughts remain in the dark. This is common enough in animal tales, but gains special significance in light of the epistemic and ethical questions raised by Agee's short story. The urge to draw from 'A Mother's Tale' a lesson in understanding the story's animals and their agony is overshadowed by the distance separating the world of those animals from humans. Nevertheless, the foremost questions raised by my reading concern not only the difficulty but the possibility of jumping some of that distance.

It could be argued that works of literature (and of art more generally) can precisely here play a notable part, in so far as they not only reflect our cultural structures of feeling (in Raymond Williams's [1977] terms) but actively participate in structuring them. As Butler writes, at stake here are 'normative schemes of intelligibility' that function 'through providing no image, no name, no narrative, so that there never was a life, and there never was a death' (Butler 2004: 146). Literature in the vein of 'A Mother's Tale' can work upon these schemes by providing not only images and names but also narratives of non-human life. All this will not do away with the anxieties inherent in the concept and practice of anthropomorphism and anthropomorphic thinking, but suggests its gradual reconfiguration. If the question of anthropomorphism is posed seriously, as a problem for thought whose solution is not predetermined by a conviction in what is proper to man and to man only, its imaginative employment may move both epistemic and ethical fences.

Note

1 I thank Dr. Robert McKay for the permission to quote from his unpublished article. McKay's talk 'Humane conditions: Animal killing and the limits of political liberalism in post-war culture' (presented 23 April 2013 at the conference *Unruly Creatures 3*, London), very similar to the article quoted, is made available by *Backdoor Broadcasting* (http://backdoorbroadcasting.net/2013/04/robert-mckay-humane-conditions-animal-killing-and-the-limits-of-political-liberalism-in-post-war-culture/).

References

Agee, J., 1968 [1952]. A Mother's Tale. *In*: R. Fitzgerald, ed. *The collected short prose of James Agee*. Boston, US: Houghton Mifflin, 221–43.

Butler, J., 2004. *Precarious life: the powers of mourning and violence*. London: Verso.

Butler, J., 2009. *Frames of war: when is life grievable?* London: Verso.

Diamond, C., 1995. *The realistic spirit: Wittgenstein, philosophy, and the mind.* Cambridge, US: Bradford Books.

Garrard, G., 2004. *Ecocriticism*. London: Routledge.

Haraway, D.J., 2008. *When species meet*. Minneapolis, US: University of Minnesota Press.

Kolozova, K., 2010. *The lived revolution: solidarity with the body in pain as the new political universal*. Skopje: Euro-Balkan Press.

McKay, R., forthcoming. James Agee's 'A mother's tale' and the biopolitics of animal life and death in post-war America. *In*: A. Hunt and S. Youngblood, eds. *Against life*. Evanston: Northwestern University Press.

Nagel, T., 1974. What is it like to be a bat? *The Philosophical Review*, 83 (4), 435–50.

Pensky, M., 2008. *The ends of solidarity: discourse theory in ethics and politics*. Albany, US: State University of New York Press.

Pollock, M.S. and Rainwater, C., 2005. Introduction. *In*: M.S. Pollock and C. Rainwater, eds. *Figuring animals: essays on animal images in art, literature, philosophy and popular culture*. Basingstoke, UK: Palgrave Macmillan, 1–17.

Potolsky, M., 2002. *Mimesis*. London: Routledge.

Scarry, E., 1985. *The body in pain: the making and unmaking of the world*. Oxford, UK: Oxford University Press.

Simons, J., 2002. *Animal rights and the politics of literary representation*. Basingstoke, UK: Palgrave Macmillan.

Sober, E., 2005. Comparative psychology meets evolutionary biology: Morgan's canon and cladistic parsimony. *In*: L. Daston and G. Mitman, eds. *Thinking with animals: new perspectives on anthropomorphism*. New York: Columbia University Press, 85–99.

Stanescu, J., 2012. Judith Butler, mourning, and the precarious lives of animals. *Hypatia*, 27 (3), 567–82.

Taylor, C., 2002. *Sympathy: a philosophical analysis*. Basingstoke, UK: Palgrave Macmillan.

Thierman, S., 2011. The vulnerability of other animals. *Journal for Critical Animal Studies*, 9 (1), 182–208.

Tyler, T., 2009. If horses had hands... *In*: T. Tyler and M. Rossini, eds. *Animal encounters*. Leiden, Netherlands: Brill, 13–26.

Williams, R., 1977. *Marxism and literature*. Oxford, UK: Oxford University Press.

13 An avian–human art?

Affective and effective relations between birdsong and poetry

Karoliina Lummaa

Diidiidii diediedie dueduedue deidaadaa,
when will come the breaking of the dawn!
When will there be peace in the skies and on the land!
Deidaadaa,
when can one sing in peace!
Deidaadaa! [...] (Lyyvuo 1946: 43)[1]

Do you know the song and the alarm call of the willow warbler (*Phylloscopus trochilus*), a green-brown coloured small bird of open woodlands? Have you ever approached its nest? If you answered yes, you might recognize its voices and resentful demeanour in the poem 'Uunilintu valittaa' ['The Willow Warbler complains'], cited above. It was written by the Finnish poet Eero Lyyvuo[2] (1904–1977) and published in 1946, in his first and only poetry collection, *Pieniä laulajia* ['Small singers']. In the first stanza,[3] the rising melody of the Willow Warbler's song is described typographically, separating notes of different heights in different lines, and creating the typical rising pattern of the song by 'lifting' the last syllables in the song transcription. This typographical patterning is probably the easiest way to recognize the willow warbler song, since bird songs are transcribed differently in different languages. The letters that Lyyvuo has used may not 'sound' right for English-speakers. In the second stanza, the Willow Warbler's call, 'fyi', is repeated: 'Fyi, fyi, fyi!' – an English version would be 'hoo-eet' or 'too-eet' (Bevis 2010: 89, 103). Again, the last letter 'i' has been raised to denote the two-note rising structure of the call.

In addition to the visual-auditory make-up of Lyyvuo's poem, there is also something bird-like in its thematics. The ill-tempered Willow Warbler is complaining about the lack of peace there is for singing. The wish for peace possibly reflects the historical context of Lyyvuo's collection: *Pieniä laulajia* was published right after the period of the three preceding wars that Finland had fought against the Soviet Union and Germany (1946). In the poem 'Metsäkiuru ihmettelee sotaa' ['Woodlark wonders about the war'] a woodlark (*Lullula arborea*) criticizes the horrible sounds made by guns and artillery. Although birds are often anthropomorphized in Lyyvuo's bird poetry, many poems share an avian view on the world, highlighting the interests and needs of birds. Thus, the second stanza of 'The Willow Warbler complains' consists of several alarm calls, 'fyi' and also 'hyi', accompanied by a few explanatory words in Finnish: 'taasko ihminen lähestyi!' ['again a man came near!']. The irritation caused by approaching humans is reflected in the second verse, where the alarm call 'fyi' is also transcribed in the form 'hyi', a Finnish expression of resentment, dislike or even disgust ('yuck' or 'ugh' in English).

Real willow warblers have definitely influenced Lyyvuo's verses. Effectively, the verses contain phonetic and rhythmic traces of their song and calls, while affectively the poem announces irritation caused by human beings. In this chapter, I examine these bird-like features of poetry. My questions are: how do bird songs and calls influence poetry and what affective consequences do these avian effects have for reading?

The connection between birdsong and poetry is certainly well-known throughout history and across cultures, but it is usually understood in terms of aesthetic inspiration or imitation (poets imitating birdsong). Moreover, in poetry and literary criticism birds are mainly given symbolic meanings belonging to the realms of human experience – birds may, for example, be emblems of freedom, spirit, immortality, or artistic superiority (Lutwack 1994; Järvinen 2000: 31–3). These ways of thinking focus on human agency and are, therefore, of no help in addressing the non-human influence in nature poetry. More fruitful ideas can be found within environmental philosophy and human–animal studies, since these fields strive to re-conceptualize the place and role of non-humans in culture. Interestingly, the question of non-human meaning-making and communication is also a hot topic currently in natural sciences. Fields as diverse as developmental ecology, neuroscience, and genetics have combined with linguistics to decipher the possible linkages between avian and human communication systems (Bolhuis and Everaert 2013).

This chapter aims to offer preliminary coordinates for understanding the complex relation between birds and poetry in a way that acknowledges the material and semiotic self-sufficiency and agential power of birds, as well as their affective presence in physical and textual environments. The chapter unfolds as follows. In the following section I conceptualize bird poetry as a naturalcultural phenomenon and describe the multicomparational method of reading. The third and fourth sections then focus on affects. While in the third section I define affect as an experiential, emotion-evoking event and trace it in Lyyvuo's poetry and in encounters between different species, before concluding, the fourth section is dedicated to avian affects and to affects evoked by species-differences and naturalcultural poetics.

Towards naturalcultural poetics

The presence of non-human interests and birdsong-like phonetic and rhythmic features in poetry leads us to challenge the idea that poetry expresses only human emotions, experiences, and ideas. Although, in many senses of the term, the language of poetry is symbolic, it is reasonable to ask whether poetry could be involved with real beings outside language and culture.

Questions concerning animals and their cultural representations, or, more generally, cultural engagements with the non-human, have stimulated a considerable amount of theoretical and critical interest. Discussions have often revolved around two specific topics, namely the inclusion of animals within a common sphere of social relations, and animals' agential role in meaning-making. The sociologist Bruno Latour (2004a: 85–6; 2010; 2011) repeatedly mentions language and fiction as devices with which non-human actors are brought inside the *collective*, i.e., a future democracy, which is constituted by associating both humans and non-humans within it. For example, talking and writing about animals as representatives of their own species and as actors affecting natural and cultural environments are ways of bringing them inside the common social reality of humans and non-humans. The feminist philosopher and historian of science Donna Haraway (1997; 2008) has written about mutually constitutive relationships between humans and non-humans and about animal representation as dependent on real animals. Haraway has introduced the term *naturecultures*, widely used nowadays, which stands for all kinds of material (bodily, physical) and semiotic (discursive, symbolic, conceptual) relations between humans and non-humans, and hence draws our attention to the fact that natural and cultural phenomena never occur in isolation from each other. Hence, my use of the term *naturalcultural poetics* is an adaptation of Haraway's naturecultures.

Levi Bryant, a philosopher who has been greatly influenced by the work of Latour and Haraway, scrutinizes the origins and mechanisms of the nature/culture distinction in a way that facilitates reassessment of the role of non-human agency in nature poetry. Bryant advocates object-oriented philosophy, a strand of thought that emphasizes ontology over epistemology and supports the ontological equality of all existing beings, be they humans, animals, plants, machines, or any other units of reality. He is a keen critic of all modes of cultural constructivism and a reluctant witness of the 'push to dissolve objects or primary substances in the acid of experience, intentionality, power, language, normativity, signs, events, relations, or processes' (Bryant 2011a: 35).

In his criticism of the nature/culture divide, Bryant makes use of the ideas of the mathematician George Spencer-Brown. In order to indicate something, we must draw a distinction between a marked space and an unmarked space, where the marked space encloses everything that can be indicated. As the unmarked remains with no such indications, it might also be said not to exist at all: for a system making a distinction, the unmarked outside remains a blind-spot. Another blind-spot is the distinction itself: once the distinction is made, we lose any account of the establishment of distinction and focus our attention on the realm of the marked – as if the distinction had never occurred. Bryant (2011a: 21) explains that distinctions between marked and unmarked create 'a reality effect where properties of the indicated seem to belong to the indicated itself rather than being effects of the distinction'. According to Bryant, contemporary philosophy and theory situate the human subject and culture inside the marked space, and non-human nature is left outside, in the unmarked space of the distinction. Following Spencer-Brown, Bryant notes that this area is deemed unknowable and without meaning. Further, as Bryant's concept of the reality effect implies, we are led to think that this distinction is completely natural and, indeed, the only possible one (ibid.).

With this philosophical gesture of shutting non-humans outside the realm of the meaningful Bryant is able to explain the more specific attitude that is common in cultural studies and art studies, i.e., the consideration of non-human animals as symbols and representations deriving their meaning solely from other cultural symbols and representations. Bryant (2011a: 21–2) continues:

> The catch is that in treating the object as what is opposed to the subject or what is other than subject, this frame of thought treats the object in terms of the subject. The object is here not an object, not an autonomous substance that exists in its own right, but rather a representation. As a consequence of this, all other entities in the world are treated only as vehicles for human contents, meanings, signs, or projections.

The challenge for future philosophy is to relocate the distinction concerning human and non-human worlds or realities:

> we must [...] redraw our distinctions in such a way as to make room for non-human objects as autonomous actors in their own right, such that these objects are not treated as merely passive screens for human projections and such that they are treated as perturbing the world in their own way. In other words, the point is to expand the domain of what can be investigated, not to limit it (Bryant 2011a: 283).

The motivation behind Bryant's claims is not only philosophical but also environmentally ethical, as he traces the philosophical roots of unsustainable action back to constructivism. A similar critique has been provided by many other theorists interested in the non-human (e.g., Latour 2004a: 32–41; Latour 2004b; Bryant 2011b; cf. Soper 1995, 4–9, 149–55).

How should we then proceed with the idea of non-humans influencing a field as strikingly *cultural* as poetry? The first step, already partly taken in the first section, is to acknowledge the status and agency of the non-humans whose influence we can detect in poetry. Bryant claims that 'the world must be a particular way for certain practices and activities like perception, experimentation, discourse, and so on, to be possible and that the world would be this way regardless of whether we perceived, experimented, or discoursed about it' (2011a: 65). In other words, the world with its materialities and non-humans comes first, and no discourse about non-humans is possible without them (see also Marshall 2002: 234; Ratamäki 2009). Obviously, birds do not participate in writing *per se* (Maran 2007; Tüür 2009). Their existence and behaviour, however, is enough for agency, since agency does not have to be intentional (e.g., Iovino and Oppermann 2014). The second step, also suggested in the quotation above from Bryant, is to acknowledge the independence of non-humans from any perceptions or discourse that we might have and make of or about them. Birds are seldom interested in us and never interested in our texts.[4]

Before taking further steps towards understanding naturalcultural poetics and defining the multicomparational method of reading, I will cite the first stanza of another bird poem by Eero Lyyvuo, called 'Metsäkirvinen esittäytyy' ['The Tree Pipit introduces itself']:

> Didididi, my name is Anthus trivialis,
> I make plants and hunters my prey.
> Iiitpryy iiitpryy, hyyi hyyi, tsiidul tsiidul, tshatsa,
> everything I will digest, I have a good stomach! (Lyyvuo 1946: 50).[5]

With this poem, we are being introduced to a small, new singer, the Tree Pipit. The tree pipit (*Anthus trivialis*) is a passerine bird that breeds in open woodlands and groves, as Lyyvuo's poem suggests. However, contrary to the poem's description, the tree pipit is not a scavenger or an herbivore but feeds mainly on insects, although sometimes it accepts plant seeds. In the second stanza, the bird comments on the kindness of its 'wife': 'Vaimoni on lempeä di, zyii zyii, hiita' ['My wife is kind di, zyii zyii, hiita'] Lyyvuo (1946: 51). Tree pipits do not form lifelong pair-bonds but they are socially monogamous, that is, they choose a steady companion for each five-month-long breeding season – so in this sense the anthropomorphic term 'wife' is fitting.

Comparing poetic description with scientifically and/or experimentally collected data is a method of reading bird poetry as naturalcultural poetics. It is worth noting that with the terms 'description' and 'data' my intention is not to evoke an opposition between fact and fiction.[6] In the context of naturalcultural poetics and the multicomparational method of reading, 'data' and 'description' simply refer to two distinctive sets of claims about birds. Data refers to what is known (through science and other modes of observation, experience, and perceptions about birds), and description refers to claims made in a particular poem or body of work under scrutiny.

Closely connected with the data/description comparison is the comparison of bird song with song transcription. In Lyyvuo's 'Tree Pipit introduces itself' the peculiar language in the first, third, fifth, seventh, and ninth verses is not birdsong, but it is not Finnish, either. It is a transcription that cannot be understood, let alone appreciated, without knowing what a tree pipit song sounds like. Comparisons can be made with songs heard live, or, perhaps preferably, using recordings of songs.[7] Despite regional and individual aberrations, the tree pipit song is always recognizable by its basic melody and rhythmic pattern. Lyyvuo has grouped certain letters and syllables to match with the phonetic and rhythmic qualities of the tree pipit song, and he uses typographical means to mediate the melodic changes: parts sung in higher notes are placed higher and vice versa. When comparing poetic song transcriptions with bird songs, other transcriptions made of the same song in the same language can be helpful. In a Finnish birdwatchers' guide specialized in representing bird songs and calls, the authors Kai Linnilä and Sari Savikko (2004: 88) give the following transcription of the tree pipit song: 'tssi tssi tssi tssi tssi tssi tsiidul tsiidul tsiidul tsiidul tsiidul hyyi hyyi hyyi hyyi si si si si si si iii'. There are two shared syllables in the transcriptions made by Lyyvuo and Linnilä and Savikko: 'hyyi' and 'tsiidul'. In addition, Linnilä's and Savikko's 'tssi' and 'si' come close to Lyyvuo's 'zyii', but Lyyvuo's 'iiitpryy', 'tshatsa', and 'diitsa' look as if they belong to another bird's repertoire.[8] However, when compared with actual tree pipit songs, Lyyvuo's spectrum of sound syllables becomes completely justified. Accordingly, the original songs of real birds are the basis on which all discourse and descriptions are constructed.

Another methodological comparison concerns song transcriptions and other verbal elements. In Lyyvuo's poem the end rhyme pairs are formed between Finnish and tree pipit syllables: 'tshatsa' – 'vatsa [stomach]', 'hiita' – 'viita [thicket]', 'hiita' – 'riita [quarrel]'. This interspecies rhyming establishes a connection between written human language (Finnish) and the transcribed tree pipit communication. The connection is reinforced if we read the poem aloud, as the avian words will be pronounced in a phonetic manner similar to that of Finnish words, with the stress placed on the first syllable of each word.

Comparisons between avian and human material-semiotic resources and influences prove the naturalcultural character of bird poetry. This implies that poetry is never really completely our own, but contains manifold non-human traces and influences. What, then, does this mean from an affective point of view? What emotions might be involved in textual and real-life encounters with birds?

Affective relations

In environmental humanities, the question of affect is usually discussed in the context of anthropogenic environmental problems. Rachel Carson's *Silent Spring* (1962) has inspired numerous scholars in their development of tools suitable for analysing affective environmental rhetoric in environmental literature and politics, green movements, and nature writing (Lutts 2000; Buell 2001; Lockwood 2012). My purpose in the following section is to discuss affect as an elemental part of reading poetry and in relation to non-humans – both of which events are often considered to be challenging or even daunting in their conceptualization.

What role does affect have in naturalcultural poetics – the writing and reading of nature poetry in the context of real-life entanglements between species? First, it is important to define the concept of affect, which has a long history in Western thinking, psychology, and philosophy. Affect is combined with feelings and emotions, as well as bodily experiences and passions or the capacity to be moved or touched by something. In the history of philosophy, affect has often been understood and studied as a personal phenomenon. During the past two decades, Continental philosophy, cultural studies, and critical theory have also addressed affect from collective and political perspectives. Thus we have witnessed the advent of 'the affective turn', which consists of diverse theoretical and methodological interests and attempts to renegotiate the roles of emotions and bodily experiences, sexual and ethnic aspects, and new information technologies in human culture (La Caze and Lloyd 2011). Moreover, the question of affect has also played an important role in theorizing human-non-human relations, as, for example, in the work of Donna Haraway, Gilles Deleuze, and Félix Guattari.

My own account of affect in this chapter is based on linguistic, psychological, and philosophical interpretations of the term. In psychology, affect is either the expression or the experience of an emotion, in the latter sense often understood as a 'first-hand reaction' in contrast with more complex cognitive functions (e.g., Moore and Isen 1990; Hogg *et al.* 2010). Following various psychological senses of the term, in linguistics affect is combined with the speaker's emotion or attitude. Speaker affects are expressed in the utterance, for example through the choice of words (Murphy 2003). Philosophical accounts of affect vary greatly, as I have already mentioned. In *Ethics* (1982 [1967]) Baruch Spinoza, the philosopher most often cited in discussions of affect, defines it as a change in one's condition, the experience of being moved and, as a result, the experience of altered power in action. By re-interpreting Spinoza's ideas, Gilles Deleuze and Félix Guattari (1987) understand affect as a change in intensity that is not tied to any particular experiencing subject. Brian Massumi (2002) elaborates the idea of prepersonal intensity by claiming that affect cannot be exhaustively expressed through language or representation.

Drawing on these diverse definitions and approaches, I understand affect as an emotionally and bodily experienced *changing event*, an intensity which can be aroused when one comes into contact with other human and non-human beings. Hence, I situate affect in physical and social encounters between humans and non-human animals, but I believe affects also emerge in reading. Poetry can evoke affects through choices of themes and words, but poetry can also be powerfully affective on the sensory and even partly unconscious levels by its use of rhythmic, visual, and auditory elements (see Kainulainen 2011).

Admiration, aesthetic pleasure, and inspiration are the first sentiments that come to mind when thinking about poets' attitudes towards singing birds and their songs. These are triggering, affective experiences that encourage further engagements and artistic exploration of the avian world (Scharfstein 1998; Tüür 2009; Bevis 2010: 16–19). While this applies to many of Lyyvuo's bird poems, he also moves beyond human aesthetic pleasure and strives for a more avian account of affects and vocalization in his poetry. In the poem 'Leppälintu häiritsijälle' ['The Common Redstart to a bully'] the bird is forcefully evicting an intruder. Here is the last stanza:

Fyitetettet, fyitetettet,
shame on you for
not already going away.
Fyitetettet,
go away!
Too-tritito! (Lyyvuo 1946: 18).

Like the willow warbler and the tree pipit, the common redstart (*Phoenicurus phoenicurus*) is a small passerine bird breeding in open woodlands and feeding on insects. It is a territorial bird and selects its nesting site from holes found in trees, tree stumps or the walls of buildings, and will occasionally also nest among rocks. The nest is constructed with the aid of grass, moss, and other vegetation, and feathers and hair are used as its lining material. As Lyyvuo's poem suggests, the common redstart will probably defend its nest and territory against avian and human intruders alike.

A strong affect is evoked in the poem. The bird first asks why the bully has approached its nest and then criticizes the making of loud noises: 'miksi pesääni lähestyit?/Hyit, hyit,/huudat, ryit!' (Lyyvuo 1946: 18) ['Why did you approach my nest?/Hyit, hyit,/you shout, ryit!']. In the second stanza the unwanted visitor is encouraged with clear resentment to leave: 'häpeä ettet/ häivy jo' (ibid.) ['shame on you for/not going away already']. However, I believe it is not only the words that evoke this forceful rejection and hostile attitude. In the first stanza, the warning call 'hyit' is repeated eight times, enhanced with the word 'ryit'. Now, 'ryit' may be an alternative transcription of the bird's alarm call, but it is also a Finnish word. The pejoratively charged verb *rykiä* means to cough or to clear one's throat very loudly. Together and through repetition the words/call transcriptions 'hyit' and 'ryit' express the bird's growing resentment. By rhyming and tonally coining 'hyit' with 'lähestyit' [come close], the poem also highlights the reason for such resentment: an approach that is rude and unwanted!

In Lyyvuo's poem nothing leads us to interpret the intruder as human. Yet this is probably the first option that comes to mind – probably because of the fourth verse 'huudat, ryit!' ['you shout, ryit!']. The verb *huutaa* means to yell or to shout and it is typically used in the context of humans. The possibility of applying this verb to non-humans is not, however, excluded. Thus, we are drawn to interpret the intruder as human, and hence the bird directing its resenting voices *at us* has a lot to do with affectivity. The Common Redstart's emotionally charged warning call 'hyit' becomes simultaneously charged with human emotions.

Above, I have defined affect as a changing event that is experienced emotionally and bodily. For humans, perceiving vocal and bodily gestures that are meant to drive one away are very affective and can provoke a range of negative emotions: shame, disappointment, sorrow, or hate. What affects are aroused by hostile rejection or driving away in birds, we do not know. What do *we* experience when we are undergoing rejection by a bird? Disappointment, or even sorrow?

Of course Lyyvuo's poem is just that, a poem. Reading poetry does affect us, however, and in the case of nature poetry we are confronted with imaginative contacts that are based on real encounters with real beings. Thus, nature poetry serves as a site for negotiating the meanings we give to non-humans, and it also helps us to mirror ourselves in relation to non-humans (see Gilcrest 2002: 36, 77–8). These negotiations are often affectively charged, as we may wish to find contact or bond ties with creatures we admire and are interested in. Birds are an especially strong case in point (Rothenberg 2005; Bevis 2010). Enriching poetry with non-human voices such as bird calls and birdsong functions catalytically in these poetic negotiations, since the voices demonstrate the weight of non-human reality, the presence of real beings prior to any discourse concerning them.

However, it would be misleading to link non-human voices with human language and poetry without noticing the friction or, indeed, the insurmountable differences between these various vocal and textual communication systems (Bevis 2010: 24). In the second stanza of the poem 'Common Redstart to a bully', Lyyvuo transcribes pieces of the Common Redstart song with the syllables 'fyitetettet' and 'too-tritito'.[9] These transcriptions seem to fit rather poorly with the actual song of the common redstart. In his afterword to the collection Lyyvuo (1946: 89) actually comments on the difficulty of using bird songs in his poetry when he writes that:

> I apologize for the violence I have had to do to bird vocalizations by for example breaking continuing trills, either with hyphens or by cutting them, in order to make pronunciation easier. Neither have I been able to avoid aberrations from the rhythms and melodies of birdsong as I have tried to gain a poetically tolerable form for the songs.[10]

Here, Lyyvuo emphasizes the primary art of the birds: the skilled vocalizations of birds come first and cannot be copied in human language and writing. Lyyvuo is also very open about his own technique – he has constantly modified his writing to match both the art of birds and his own poetic standards.

Instead of thinking of his poetry in terms of imitation, Lyyvuo seems to appreciate the original subjecthood and artistry among birds that humans cannot ever really imitate in their speech, writing or music. Lyyvuo, like other poets interested in birdsong, creates something that cannot be placed solely inside or outside culture – if we still want to talk about poetry in these terms. Poetry employing birdsong could perhaps be characterized as the combined art of skilled human and non-human subjects. It should not be considered as mimesis since mimesis is not possible, and it is not representation, either. Representation indicates a move from one realm to another, in this case from the realm of not-yet-meaningful nature to the realm of meaning-making (or meaning-establishing) culture (Morton 2007: 47–54; Bryant 2011a: 20–4, 134). In this line of thought, the instinct-based background noise of birdsong is transformed through poetic means to fit within culture, in a poem expressing some human idea or content. As bird poetry is an instance of naturalcultural poetics, I understand it differently, that is, as the combined art of, on the one hand, human and avian forms (rhythms, sounds, signs) and, on the other, contents (affects, experiences, interests).

But if bird poetry is an art of combination rather than representation, how much 'bird' is really present in bird poetry, and what affects are evoked in avian–human poetics?

Affects and the unknown

With Lyyvuo's poem 'Kolmivarpainen tikka pöyhkeänä' ['Cocky Three-Toed Woodpecker'], we are introduced both visually and auditorily to two different types of sounds that all woodpeckers make: calls and pecking/drumming. The call of the Three-Toed Woodpecker (*Picoides tridactylus*) is transcribed using two different syllables, 'ki' and 'ti', which are divided into two different lines: 'ki' always appears higher and 'ti' lower (Lyyvuo 1946: 63–4). This typographical pattern once again corresponds to the two different tones in the woodpecker's call: the higher and louder one, and the lower and quieter one.[11] These calls have several diverse social meanings, ranging from calling mates to warnings of intruders.

The functions of drumming or pecking are also varied. Slower and softer pecking noises come from the bird's efforts to find food in tree bark and to excavate nesting holes, whereas the very fast and hard drumming is an elemental part of a woodpecker's sexual display. The repeated 'pretereteret' in Lyyvuo's poem is obviously drumming, as the poem's theme implies. Drumming testifies male fitness and is meant to attract the opposite sex, as well as to inform other woodpeckers of the same sex. The males usually choose noise-enhancing surfaces for their drumming, such as very hard wood or even metal. The drumming starts very loud and fast, and fades in volume and speed towards the end of each set. Lyyvuo demonstrates this change by positioning the syllables 'Pre-te-re-te-ret' in a descending manner. Hence, the descending position of the syllables does not function in the same way as in transcriptions of the two calls of the three-toed woodpecker or of bird songs more generally: the descent is not tonal or melodic but refers mainly to volume.

Drumming on noise-enhancing surfaces has the same function as song has for songbirds, comparable with the display of colourful plumage for many bird species: they are all means to attract the female and to show off the male's abilities. In Lyyvuo's (1946: 63) poem, the Woodpecker is also boasting about its feathers: 'mulla hienot on höyhenet!/Pretereteret!/Kitikitikiti – kitikitikiti,/silkinmusta ja lumenviti!' ['I have such fine feathers!/Pretereteret!/Kitikitikiti – kitikitikiti,/silk black and snow white!']. The contrasting colours of deep black and bright white appear to flicker in turn in the beholder's eyes: 'vuorotellen ne silmään vilkkuu' (Lyyvuo 1946: 64) ['in turn they shine']. The feathers are also characterized as 'aatelishöyhenet' ['noble feathers'], which creates an implicit allusion to high social rank and extensive wealth. In the course of evolution any markers of plentiful breeding conditions and good physical health have been crucial for females, as they maximize the quantity and quality of offspring (Collins 2012).

Read in this way, male pride regarding noble feathers is not anthropomorphic at all – it is an avian message conveyed in human language. How conscious these activities of sexual display are, is another question. Some researchers are ready to claim that there is a certain amount of enjoyment involved in the acts of display and pair-bonding (Kroodsma 2005: 76, 201, 276; Birkhead 2013: 196–202). What interests me here most is the presence of an emotion that is genuinely avian. Lyyvuo's whole poem embodies the pride and potency of the male Three-Toed Woodpecker who is advertising himself to females. It is a strongly sexual emotion that we might not feel at ease to share or even understand, despite the obvious fact that humans as mammals also have their own naturalcultural means of sexual display.

There is a vast and perhaps insuperable gap between human experience and avian experience, resulting from the anatomical, physiological, psychological, social, and environmental differences between humans and birds, and also between different bird species. What poetry gives us is a hint of emotions belonging to those rich avian worlds that are very other to our own world. Poetry cannot imitate or represent avian emotions, just as it cannot imitate or represent their songs. But poems can open a space where we can experiment with our words, signs, and empathy in order to approach meanings and feelings that belong in the avian world. In this sense nature poetry is highly affective: it engenders bodily and emotionally changing events that may promise to lead us beyond species-differences. Or, in Deleuze and Guattari's (1987) terms, bird poetry creates an event of becoming-animal, an uplifting moment of experiencing powers, drives, or desires that belong to the realm of the avian. This, it must be admitted, is an optimistic interpretation of bird poetry and its strange avian sounds and experiences.

A more pessimistic interpretation can be traced, for example, in John Bevis's (2010) account of transcriptions of bird calls and songs, which he calls 'bird words', in his book *Aaaaw to zzzzzd: The Words of Birds*. He characterizes bird words as 'bizarre, nonsensical, sometimes pretty, sometimes comic'. In another context he writes: 'The sounds of the singing are just that, sounds without semantics, making what is expressed seem to me unhindered in its rawness and potency' (Bevis 2010: 15). And he continues: 'We cannot ground the emotions we hear in any birdland rationality. We cannot compare how the songs of the nightingale or mockingbird affect us with how they affect another of the same species, fluent in the same language' (ibid.). Bevis's emphasis on the incomprehensibility of avian messages and affects leads us to an important problem that has been implicitly present all along: naturalcultural poetics makes our own language *strange* by allowing non-human voices, affects, and other influences stream through the verses (see also Morton 2013: 110–15, 142).

If we cannot totally understand the nonsensical elements induced by bird words and if we cannot really tell whether our descriptions of avian experiences can be shared or not, does this not evoke a whole new series of affects – more unpleasant ones? In addition to the awkward, disappointed, or sad emotions aroused by Lyyvuo's poems about avian hostility and suspicion towards our own species, these affective experiences of not understanding and not sharing seem to be another more implicit yet meaningful element in Lyyvuo's bird poetry.

Lyyvuo's poetry, and naturalcultural poetics more generally, seem to evoke both almost ecstatic affective experiences of sharing and also discouraging feelings of difference and distance. Levi Bryant (2011b) balances between these two affective extremes in his essay 'Wilderness Ontology', where he presents a sharp criticism of the idea of humans as active and sovereign beings who give meaning to passive, non-signifying nature. Wilderness ontology is Bryant's attempt to think of a shared naturalcultural world of humans and non-humans as a thoroughly communicative reality that is not solely governed or understood by humans. The name wilderness ontology refers to the way in which humans experience an environment where all human mastery is lost:

> in the wilderness I find myself regarded by beings other than humans – the wolves, bears, birds, and so on – and in a field of languages and signs that I scarcely understand. What does the howl I hear off yonder signify? Should I be alarmed by the hoot of that owl? Why did the forest suddenly grow quiet? What caused that branch to snap? Are those approaching clouds a danger or gift? Was this trail created by humans or deer? What are those birds talking about in their songs? In the wilderness I am no longer a sovereign or master, but a being among other beings. In short, in the wilds we encounter other beings as both agencies and as entities with which we must negotiate (Bryant 2011b: 21).

Although Bryant describes the loss of human sovereignty, he still affirms the possibility for humans to understand non-human languages up to the point of inevitable interspecies communication. Bryant's optimism is valuable when we consider the possibilities of studying naturalcultural poetics and its various affective consequences. We may not know what the birds know, or communicate with the same language, or experience exactly the same emotions, but as bodily and semiotic creatures living in close approximation with one another, our messages, interests, and affects occasionally overlap in life and in writing.

Notes

1 I have made all translations of Lyyvuo's poetry with a focus on the content of each poem. Thus I have concentrated on the meanings of words and expressions and have not tried to find phonetic and rhythmic equivalence between my translations and the original text. In consequence, I have preserved the original Finnish transcriptions of birdsongs in my translations of Lyyvuo's poems.
2 Lyyvuo is practically forgotten by the literary historians and researchers. Although Aaro Hellaakoski, one of the most praised and loved nature poets in Finland, encouraged Lyyvuo to write this collection, which was published by one of the major publishing houses in Finland, Lyyvuo remained unmentioned in national literary histories until the publication of Juri Nummelin's (2007) book *Unohdetut kirjailijat* ['Forgotten authors']. In the 1940s, one review of Lyyvuo's collection *Pieniä laulajia* was published in the natural-scientific journal *Luonnon tutkija* ['Researcher of nature']. That Lyyvuo is not well known can of course be explained in many ways, but it appears to have much to do with the exceptionality of his poetry. He used transcriptions of bird calls and songs in his bird poetry, and he was one of the first poets to write about birds as animals, not just as artistic, religious or patriotic symbols, or metaphors for subjective human experiences.
3 In this chapter, the term 'verse' refers to a single line of poetry, and 'stanza' refers to a structured group of verses. I also use the term 'line' to indicate the placing of individual syllables in a single verse (see the example of Lyyvuo's poem in endnote 5, below).
4 Music, however, is a totally different issue (see West *et al.* 2012 [2004]).

5 Here is a scanned photograph of the poem:

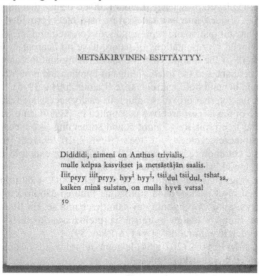

METSÄKIRVINEN ESITTÄYTYY.

Dididi, nimeni on Anthus trivialis,
mulle kelpaa kasvikset ja metsästäjän saalis.
Iiit$_{pryy}$ iiit$_{pryy}$, hyy^i hyy^i, tsii$_{dul}$ tsii$_{dul}$, tshat$_{sa}$,
kaiken minä sulatan, on mulla hyvä vatsal
50

Figure 13.1 Eero Lyyvuo, 'Metsäkirvinen esittäytyy' ['The Tree Pipit introduces itself'], Lines 1–4

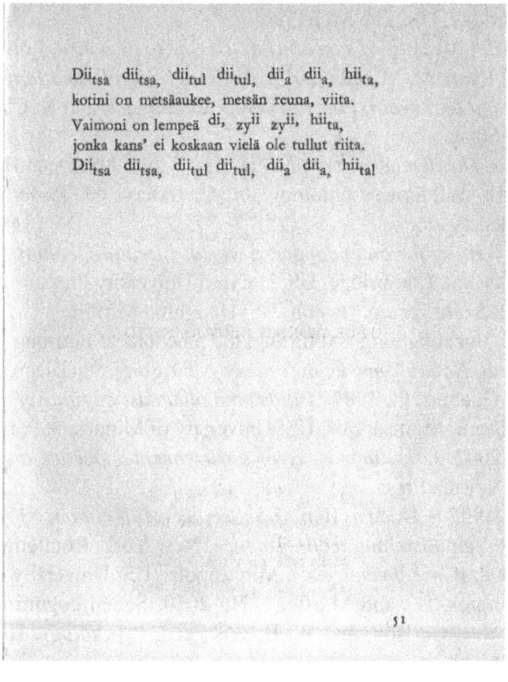

Dii$_{tsa}$ dii$_{tsa}$, dii$_{tul}$ dii$_{tul}$, dii$_a$ dii$_a$, hii$_{ta}$,
kotini on metsäaukee, metsän reuna, viita.
Vaimoni on lempeä di, zy^{ii} zy^{ii}, hii$_{ta}$,
jonka kans' ei koskaan vielä ole tullut riita.
Dii$_{tsa}$ dii$_{tsa}$, dii$_{tul}$ dii$_{tul}$, dii$_a$ dii$_a$, hii$_{ta}$l

51

Figure 13.2 Eero Lyyvuo, 'Metsäkirvinen esittäytyy' ['The Tree Pipit introduces itself'], Lines 5–9

6 This opposition has been illuminatingly analysed by Bruno Latour. In his criticism of the superior role played by science, Latour (2004a) claims that all talking and thinking about nature is divided into two categories: hard facts (provided and communicated solely by scientists) and social representations (created and shared by the rest of the population). Social representations of nature have no bearing in decision-making. In this situation, all of the complexities of social reality and the complexities of natural reality are kept apart, and the intertwinings of humans and non-humans remain largely out of sight and beyond political relevance (Latour 2004a: 32–5).

7 For example, the Internet provides numerous easily accessible resources of bird songs and calls. One of the largest archives is supplied by Xeno-Canto, at xeno-canto.org.

8 In English, the tree pipit song might sound something like 'seea-seea-seea', 'srihb', 'sip', 'sipsipsipteezeteezeteeze', or 'teez' (Bevis 2010: 99, 100, 102).

9 In English, the common redstart song is transcribed, for example, as 'hooeet', 'hweet-tuc-tuc', and 'tooick' (Bevis 2010: 89, 90, 103).

10 The Finnish original is as follows: 'Pahoittelen sitä, että minun on lausumisen helpottamiseksi ollut pakko tehdä väkivaltaa lintujen äänille katkomalla esimerkiksi jatkuva liverrys, joko väliviivoilla, tai paloittelemalla se. En liioin ole voinut välttyä poikkeamasta lintujen laulun rytmistä ja melodiasta koettaessani saada lauluille runollisesti siedettävän muodon' (Lyyvuo 1946: 89).

11 Interestingly, John Bevis's (2010: 64) lexicon of bird words has only one transcription for the three-toed woodpecker: 'pik'.

References

Bevis, J., 2010. *Aaaaw to zzzzzd: the words of birds: North America, Britain, and northern Europe*. Cambridge, US: The MIT Press.

Birkhead, T., 2013 [2012]. *Bird sense: what it's like to be a bird*. London: Bloomsbury.

Bolhuis, J.J. and Everaert, M., eds. 2013. *Birdsong, speech, and language: exploring the evolution of mind and brain*. Foreword by R.C. Berwick and N. Chomsky. Cambridge, US: The MIT Press.

Bryant, L., 2011a. *The democracy of objects*. Ann Arbor, US: Open Humanities Press.

Bryant, L., 2011b. Wilderness ontology. *In*: C. Jeffery, ed. *Preternatural*. New York: Punctum Books, 19–26.

Buell, L., 2001. *Writing for an endangered world: literature, culture, and environment in the U.S. and beyond*. Cambridge, US: Harvard University Press.

Carson, R., 1962. *Silent spring*. Boston, US: Houghton Mifflin.

Collins, S., 2012. Vocal fighting and flirting: the functions of birdsong. *In*: P. Marler and H. Slabbekoorn, eds. *Nature's music: the science of birdsong*. San Diego, US: Elsevier, 39–79.

Deleuze, G. and Guattari, F., 1987. *A thousand plateaus: capitalism and schizophrenia 2*. Trans. B. Massumi. Minneapolis, US: University of Minnesota Press.

Gilcrest, D.W., 2002. *Greening the lyre: environmental poetics and ethics*. Reno, US: University of Nevada Press.

Haraway, D., 1997. *Modest_Witness@Second_Millennium: FemaleMan©_meets_ OncoMouse™: feminism and technoscience*. New York: Routledge.

Haraway, D., 2008. *When species meet*. Minneapolis, US: University of Minnesota Press.

Hogg, M.A., Abrams, D., and Martin, G.N., 2010. Social cognition and attitudes. *In*: G.N. Martin, N.R. Carlson, and W. Buskist, eds. *Psychology*. Harlow, UK: Pearson Education, 646–77.

Iovino, S. and Oppermann, S., 2014. Introduction: stories come to matter. *In*: S. Iovino and S. Oppermann, eds. *Material ecocriticism*. Bloomington, US: Indiana University Press, 1–17.

Järvinen, A., 2000. *Ihmiset ja eläimet: 'humanistin eläinkirja'*. Helsinki: WSOY.

Kainulainen, S., 2011. *Kun sanat eivät riitä: rytmi, modernismi ja Eila Kivikk'ahon runous*. Turku, Finland: Turun yliopisto.

Kroodsma, D., 2005. *The singing life of birds: the art and science of listening to birdsong*. Drawings by N. Haver. Boston, US: Houghton Mifflin Company.

La Caze, M. and Lloyd, H.M., 2011. Editor's introduction: philosophy and the 'affective turn'. *Parrhesia*, 13, 1–13.

Latour, B., 2004a. *Politics of nature: how to bring the sciences into democracy*. Trans. C. Porter. Cambridge, US: Harvard University Press.

Latour, B., 2004b. Why has critique run out of steam? From matters of fact to matters of concern. *Critical Inquiry*, 30, 225–48.

Latour, B., 2010. An attempt at a 'Compositionist Manifesto'. *New Literary History*, 41, 471–90.

Latour, B., 2011. Waiting for Gaia: composing the common world through arts and politics. A lecture at the French Institute, London, November 2011 for the launching of SPEAP (the Sciences Po program in arts and politics). Available from: www.bruno-latour.fr/node/446. [Accessed: 1 February 2015].

Linnilä, K. and Savikko, S., 2004. *Onko lintu kotona? Sata siivekästä*. Helsinki: Teos.

Lockwood, A., 2012. The affective legacy of *Silent Spring*. *Environmental Humanities*, 1, 123–40.

Lutts, R., 2000. Chemical fallout: *Silent Spring*, radioactive fallout, and the environmental movement. *In*: C. Waddell, ed. *And no birds sing: rhetorical analyses of Rachel Carson's Silent Spring*. Carbondale, US: Southern Illinois University Press, 17–41.

Lutwack, L., 1994. *Birds in literature*. Gainesville, US: University Press of Florida.

Lyyvuo, E., 1946. *Pieniä laulajia: linturunoja*. Helsinki: WSOY.

Maran, T., 2007. Towards an integrated methodology of ecosemiotics: the concept of nature-text. *Sign Systems Studies*, 35 (1/2), 269–94.

Marshall, A., 2002. *The unity of nature: wholeness and disintegration in ecology and science*. London: Imperial College Press.

Massumi, B., 2002. *Parables for the virtual: movement, affect, sensation*. Durham, US: Duke University Press.

Moore, B.S. and Isen, A.M., eds. 1990. *Affect and social behavior*. New York: Cambridge University Press.

Morton, T., 2007. *Ecology without nature: rethinking environmental aesthetics*. Cambridge, US: Harvard University Press.

Morton, T., 2013. *Realist magic: objects, ontology, causality*. Ann Arbor, US: Open Humanities Press.

Murphy, M.L., 2003. *Semantic relations and the lexicon*. Cambridge, UK: Cambridge University Press.

Nummelin, J., 2007. *Unohdetut kirjailijat*. Helsinki: BTJ Kustannus.

Ratamäki, O., 2009. Luonto, kulttuuri ja yhteiskunta osana ihmisen ja eläimen suhdetta. *In*: P. Kainulainen and Y. Sepänmaa, eds. *Ihmisten eläinkirja: muuttuva eläinkulttuuri*. Helsinki: Palmenia, 37–51.

Rothenberg, D., 2005. *Why birds sing: a journey through the mystery of bird song*. New York: Basic Books.

Scharfstein, B.-A., 1998. Animals as artists. *In*: M. Seppälä, J.-P. Vanhala, and L. Weintraub, eds. *Animal: Anima: Animus*. Pori, Finland: Pori Art Museum Publications and FRAME Publications, 162–80.

Soper, K., 1995. *What is nature? Culture, politics and the non-human*. Oxford, UK: Blackwell.

Spinoza, B. 1982 [1967]. *Ethics and selected letters*. Indianapolis, US: Hackett.

Tüür, K., 2009. Bird sounds in nature writing: human perspective on animal communication. *Sign Systems Studies*, 37 (3/4), 580–613.

West, M.J., King, A.P., and Goldstein, M.H., 2012 [2004]. Singing, socializing, and the music effect. *In*: P. Marler and H. Slabbekoorn, eds. *Nature's music: the science of birdsong.* San Diego, US: Elsevier, 374–87.

PART IV
Methodological afterword

14 Ethnographic research in a changing cultural landscape

Karen Dalke and Harry Wels

Much has been written about increasing scientific evidence that can no longer be negated, showing that distinctions between human and non-human animals are not so much in kind as in degree. Distinctions between humans offered, like gender, 'race', (dis)ability), and humans and non-human animals – 'human–animal distinctions' – suggested and constructed in science, popular media, and religion, can all be unpacked as socio-cultural and socio-political constructions, unveiling more about the ideological, religious, gender, and/or species stance of the people who formulate them than about the social realities and interactions they try to understand (cf. Bekoff 2007; Derrida 2008; Haraway 2008; Bourke 2011; King 2013). It has come to a point where we can perhaps follow Calarco (2008: 149), who states on the last page of his book, explicitly following Haraway: 'We could simply let the human–animal distinction go'.

Humans and animals have a long history of interaction. James Serpell's (1986: 4) *In the Company of Animals: A Study of Human–Animal Relationships* provides a timeline beginning with domestic wolves in the Near East 14,000–12,000 years ago, closely followed by domestic sheep and goats. This human–animal interaction continues to include domestic cattle around 9,000 years ago followed by horses, asses, camels, and water buffalo. Around 3,000 to 4,000 years ago the domestic cat emerged in Egypt and continues to be a primary pet in many cultures today. Understanding these relationships, who we are, and how we interact with the environment over time provides the perfect context for anthropological inquiry.

Anthropology is the study of culture. This holistic discipline reveals data through relationships of and *with* the observed. We must be forever vigilant recognizing that how animals are viewed is contextualized in a culture influenced by historical roots. In *Regarding Animals* Arluke and Sanders (1996) state that anthropology has recognized animals in its ethnographic research. What they were referring to was ethnographic research conducted by anthropologists such as E.E. Evans-Pritchard (1956) and Clifford Geertz (1973). These studies first observed that animals serve as useful instruments of culture because they are highly flexible symbols. However, the relationship was viewed through an anthropocentric lens.

Early anthropology had its roots in colonialism impacted by theoretical constructs, such as unilineal evolution. This progressive unilineal perspective categorized humans in an orderly fashion from primitive to fully enculturated. Although anthropologists recognized the interconnectedness, it is not surprising that non-human animals were viewed as something outside 'civilized culture'. Structural functionalism provided a way to understand human cultures where animals often played a central role.

In the 1930s, Evans-Pritchard began studying the political and economic systems of the Nuer to benefit British expansion and interest in what is now South Sudan. His early writings focus on the intermingling of cattle with human cultural rituals and everyday life. The reader becomes immediately aware of the significance of cattle, but from an anthropocentric perspective. Much about the Nuer culture and people has changed since the time Evans-Pritchard studied them in 1930. In the 1930s Nuer culture was centred on cattle and few people understood the concept of currency (Hutchinson 1996: 296). By 1980, the cattle were no longer considered as important as they once were. Cash and cattle had become interchangeable and most people found it convenient to use cash when paying taxes and fees and exchanging grain or fish. Instead of being the main means of exchange, cattle were used for more important things such as marriage and sacrifice (Hutchinson 1996: 63). Although a central focus of the ethnographic research, the agency of cattle was not a focus, rather they were a means for human functioning. The focus on animals changes as new theoretical approaches emerge.

Geertz (1973: 412–54) in the chapter *Deep Play: Notes on the Balinese Cockfight* moves beyond the structural analysis of Evans-Pritchard and looks to symbolism where the cockfight embodies the culture, social networks, and rituals of Balinese life. This study is cited frequently for its detail and analysis. On the surface it seems to highlight the 'natural' behaviour of the cocks, but with further examination it becomes evident that this behaviour is anything but natural. These animals are bred for aggression and have no way to escape their opponent. Although it emphasizes the importance of animal symbols it perpetuates the Cartesian dualism of nature and culture (Bamberg 2007). Studies like this have animals in them, but they do not examine the experience with them that Haraway advocates with multispecies ethnography centring on how a multitude of organisms' livelihoods shape and are shaped by political, economic, and cultural forces (qtd in Kirksey and Helmreich 2010).

Rather than provide a review of every anthropological study involving animals over time, Evans-Pritchard and Geertz show how animals have been and continue to be a central focus of anthropological inquiry, but the questions change as theory develops. As anthropologists, we explore culture through the lens of the era. We are part of the holism, which influences our gaze. As the field of human–animal studies produces research on animal agency, our theoretical constructs must adapt. Participant observation, a hallmark of anthropological inquiry, provides a method for inquiry as it has in the past, revealing information that would be difficult to gather in any other way. Anthropology has a long history of dealing with ethnocentrism and is positioned to confront a species-centric gaze. However, recognizing the agency of animals creates new challenges. How do we ask questions when language is limited that grants equality with other species? How do we know we are studying the agency of the animal and not another layer of human interpretation? Theoretically we can agree with abandoning an anthropocentric lens and include non-human animals in our research, but to do so we must confront the practical methodological challenges that ask for research methodologies without words.

The question here is of course how one may be able to 'interview' animals, if they do not have a language that we share, understand, or are able to speak? In other words, how do we read their social behaviour and communication if we cannot make use of language? This refers to the research-methodological implications and the consequences of this shift in our thinking about researching the social configurations in and between humans and human and non-human animals. Humans are usually studied by using research methodologies relying almost solely on words/texts/language/*logos* (both in the quantitative and qualitative approaches to social scientific research),[1] whereas research into animals is done through experiments and various types of observations, translating the results into words (i.e., scientific texts), but not using words themselves in researching the non-human animals. In the social sciences this methodological approach has led to a logo-centrism that makes it difficult to extend the research beyond the human species, let alone that it allows for researching human–non-human configurations with one single trans-species qualitative methodological approach, taking the agency of all included in the research into proper and non-hierarchical, non-speciesist account (cf. Buller 2014: 314). Therefore the question is: what can ethnography and ethology learn from each other, what can they combine, and how can they complement each other, in order to start to develop a qualitative trans-species 'research methodology without words?' In order to develop this methodology, to which Buller (ibid.) refers as 'the experimental movement', we want to focus in particular on the qualitative domain of researching human and non-human animals in the ethological fieldwork traditions of Jane Goodall and others, as well as on the ethnographic fieldwork traditions in social scientific and anthropological research. In order to be able to combine ethological and ethnographic research methodologies, we will narrow the latter down to ethnographic research among humans that, just like in the ethological traditions we refer to, does not make use of words or spoken language. This is building further on what Lestel, Brunois, and Gaunet (2006: 168) propose to call 'ethno-ethology' that 'grants all living beings the status of relational beings, that is, agents interacting on the phenomenon of "culture" that was hitherto reserved for human beings'. According to Lestel, Brunois, and Gaunet (ibid.), this approach should be complemented with 'etho-ethnology', in which animals will 'be defined as a natural or artificial, human or non-human agent that attempts to control its actions and those of others as a result of the significations it ascribes to their behaviors'. Taken together this approach would be able to study the 'shared lives' of human and animals, 'on the paradigm of convergences between the two, of life shared by intentional agents belonging to different species' (Lestel *et al.* 2006: 156). Looking for methodologies without words, for instance, has led us to disability studies where there is a research tradition trying to understand the life worlds of, for instance, deaf/blind mutes; we have come across methodologies without words and we have located research methodologies that take the concept of empathy as their point of departure, which we will explore in more detail below.

Empathy in ethological and ethnographic fieldwork[2]

A great breakthrough in the study of social animals, certainly in the popular imagery, was brought about by the ethological fieldwork of Jane Goodall on chimpanzees, Dian Fossey on mountain gorillas, and Biruté Galdikas on orangutans (Montgomery 2009). It was considered revolutionary because '[t]heir methods of study were much more like those approved for anthropologists than like those approved for wildlife biologists' (Marshall Thomas 2009: xiv). It was also seen as groundbreaking because with their particular fieldwork approach they challenged 'the masculine world of Western science' (Montgomery 2009: xix), and their work was therefore to be heralded as a triumph of the feminine approach to science (Montgomery 2009: 238). An approach that could be captured by the Japanese word *kyokan*, derived from Kawai Masao, a primate researcher from Japan. '*Kyokan* means *becoming* fused with the monkeys' lives where, through an intuitive channel, feelings are mutually exchanged', that is, 'to "feel one" with them' in a shamanistic, (almost) spiritual, way (Montgomery 2009: 238–9; emphasis added). A shamanistic way of 'becoming animal' is also proposed by Deleuze and Guattari (1980): 'A similar ontological process to shamanism which undermines fixed identities, as well as crossing thresholds' (qtd in Woodward 2008: 4; also see the extensive discussion on 'becoming' in Ten Bos 2008: 75–91). This view is described in a tongue-in-cheek manner by Michael Ryan (2010: 478; emphasis added) writing about Merlin Tuttle's interest in bats, saying that he had 'rarely met anyone with such *a feel* for their study animal', and jokingly remarking he thought that 'Merlin was part bat' (Ryan 2010: 477). At the same time it should be recognized that this close proximity to, and scientific study of, animals can also lead to a rather mechanistic perspective on animals, as the life and work of famous and Nobel-prize-winning ethologist Niko Tinbergen illustrates. He is regarded as the father of the strand of ethology that assumes that 'we cannot know what an animal feels or what it intends, so scientists should not speculate on its subjective experience' (Kruuk 2003: 3). In the same biography though, it is observed that this may be attributed to the fact that:

> Niko never felt comfortable in the presence of dogs, and did not keep any himself in later life. It is quite possible that if he had grown up with one, it would have been more difficult for him to see animals mechanistically as he did in his later science, and it would have been more difficult to sideline animal emotions and feelings (Kruuk 2003: 19).

One of the best-known scientists studying empathy in social animals, in his case chimpanzees, is probably Dutch primatologist Frans de Waal (2008: 281), who defines empathy as the capacity to:

a) be affected by and share the emotional state of another;
b) assess the reasons for the other's state;
c) identify with the other, adopting his or her perspective.

This definition seems to refer as much to the object of study, that is, empathy in chimpanzees, as to empathy as a research methodology, without de Waal making the distinction explicit. It becomes clear from this example that we need to make a distinction between researching empathy in social animals, and using empathy as a research methodology. Our suspicion is that many scientists interested in empathy in animals actually also use empathy as a 'research methodology without words', without overtly stating so, and without defining clearly what empathy means as a research methodology. De Waal's approach actually seems to be very similar to the definition of *kyokan* given above. This should not come as a surprise, as throughout his many books de Waal has always written very highly of his Japanese colleagues, who, according to him, were also the first to start researching 'culture' among primates. This research was inspired by the work of Kinji Imanishi, who argued that non-human animals are not instinct-driven, i.e., Cartesian machines, and do not learn and know by instinct (i.e., genetically), but through social learning (i.e., culture) (de Waal 2001). This has resulted in various culture studies among animals, led by Japanese researchers (ibid.). De Waal's appreciation for the specific contribution of Japanese primate researchers is shared by others, Asquith (2002) among others. In the foreword of his 2002 book, Hiroyuki Takasaki, one of the translators of the work of Imanishi, writes that the 'discovery of cultural behaviors [of primates] is also traceable to his worldview, which encourages anthropomorphism when judged appropriately' (Takasaki 2002: xii).[3] It seems that de Waal's interaction with his Japanese colleagues over the years has almost inevitably led him to the concept of empathy, to which he has devoted one of his most recent books (2009). De Waal is interested in the extent to which empathy can be observed in non-human animals and, more specifically, primates. He argues convincingly, as others do (see for instance Bekoff 2007; 2013), that empathy is widespread among social animals, but he does not seem to be explicitly aware of his use of the concept as a research methodology.

Empathy can be philosophically framed as 'becoming other', a concept most explicitly developed by Giles Deleuze and Félix Guattari (1980), while at the same time forming part of a broader discourse in postmodern (primarily French) philosophy that seeks to capture and conceptualize the volatility of social reality. Deleuze and Guattari participate in this discussion by introducing the concept of 'becoming', which stands for the ultimate fluidity and flux of social reality, a reality that never reaches any final state or destination; 'becoming' never totally 'becomes', as it always remains an exploration of the other (Janssens and Steyaert 2001: 131). In that sense 'becoming' always lingers 'in the middle of difference' (Ten Bos 2008: 80). 'Becoming animal' is therefore not an attempt to ultimately become the animal itself, but to try and understand the animal from the middle of one's relation with it; from the 'middle of difference' (ibid.). Becoming aims to avoid looking at the Other from a dominant position of the self. Describing animals as 'non-human animals' for instance, categorizes them as something that is not similar to the dominant human, instead of trying to approach the animal from its own self (cf. Neolan in Janssens and Steyaert 2001: 128). Becoming steers clear of using difference as an absence, or failure, of similarity, but aims instead at studying the other from the perspective of difference itself, from the middle. This means that becoming is a process of anonymizing the human subject, trying to reach the middle of difference in the relation with the other. In a way, becoming leaves the self behind in its exploration of the other; in its journey to the middle. Interestingly, Deleuze and Guattari refer to children as an example of how to relate to animals without taking the self, which is not yet developed as such in children, along (in Ten Bos 2008: 89). This brings to mind the Biblical notion that to 'really' believe in God, is to *become* like a child, as only they will enter the Kingdom of God (cf. Matthew 18: 3). This parallel is even more suggestive as Ten Bos (2008: 87), following George Kampis, suggests that becoming basically asks for a 'knowing without knowing', which seems to echo the Biblical dictum that 'faith is the substance of things hoped for, the evidence of things not seen' (Hebrews 11: 1; King James Version). Becoming asks for 'intensities of relationship', not with a single animal, because in order to try and understand the animal we must appreciate the totality or *collective of contexts*[4] in which the animal lives, which in itself is continually becoming (cf. Ten Bos 2008: 87–9). 'Intensive relationships' with the other, be it animals, plants, or other, facilitate becoming. Barbara McClintock, the famous geneticist, basically asked of her students to work towards 'becoming plant', without the vocabulary of Deleuze and Guattari yet developed or available to her at the time. The students had to stay with a maize plant when it germinated and grew, cell by cell, into a full-grown maize plant, and she herself did the same. As McClintock said, 'I don't really know the story if I don't watch the plant all the way along' (qtd in Fox Keller 1983: 198). Only by cultivating an intense relation with the plants' becoming was she able to understand them and to work on their genetic modification:

> Over and over again, she tells us one must have the time to look, the patience to 'hear what the material has to say to you', the openness to 'let it come to you'. In short, one must have 'a feeling for the organism' (Fox Keller 1983: 198).

McClintock maintains that '[g]ood science cannot proceed without a deep emotional investment on the part of the scientist' (Fox Keller 1983: 198). In so doing she sounds slightly like the trackers described above: '[A]t any time, for any plant, one who has sufficient patience and interest can see the *myriad signs* of life that a casual eye misses' (Fox Keller 1983: 200; emphasis added). A research methodology without words. As explained above, Deleuze and Guattari offer an interesting avenue to develop a research methodology without words (Wels 2013). Donna Haraway (2008) nevertheless criticizes Deleuze and Guattari's conceptualization of becoming in quite strong words. We want to pay attention to why Haraway is so openly and explicitly annoyed with Deleuze and Guattari, in order to understand her answer to them; an answer that seems most fitting to the kind of fieldwork-oriented research methodology without words that we want to develop in this chapter. Haraway (2008: 28) seems particularly offended by the following formulation by Deleuze and Guattari:

> Bands, human or animal, proliferate by contagion, epidemics, battlefields, and catastrophes. [...] All we are saying is that animals are packs, and packs form, develop, and are transformed by contagion. [...] Wherever there is multiplicity, you will find also an exceptional individual, and it is with that individual that an alliance must be made in order to become animal.

Directly following this quote, Haraway (2008: 28) writes: '[t]his is philosophy of the sublime, not the earthly, not the mud'. Because of this orientation towards the sublime, she accuses them of a 'disdain for the daily, the ordinary' (Haraway 2008: 29) and therefore they do not have 'the courage to look [...] such a dog in the eye', so that readers of their book 'will learn nothing about actual [animals] in all this' (ibid.). It leads Haraway (2008: 30) to write: 'I am not sure I can find in philosophy a clearer display of misogyny, [...] incuriosity about animals, and horror at the ordinariness of flesh'.

On the basis of this critique Haraway (2008) presents her own conceptualization of 'becoming *with*' which does not so much come from the middle of a relationship like Deleuze and Guattari are proposing, but takes 'the encounter', in a Levinasian way,[5] as crucial. It is an encounter between species, a 'becoming with' without words, an embodied responsiveness, a 'worlding' in which '[a]nimals are [...] full partners' (Haraway 2008: 301); '"becoming with" is "becoming worldly"' (Haraway 2008: 287). Haraway considers this worlding crucial in the encounter. She bases this to a large extent on the work of South African ethologist Barbara Smuts and her research among baboons in Kenya. In order to be 'in the mud' and as close as possible to the baboons, Smuts tried to do that the usual way, i.e., 'hanging out' *around* them, hoping that at some stage of her research the animals would forget about her and let her do her research *among* them, but that did not happen. On the contrary, 'They frequently looked at her, and the more she ignored their looks, the less satisfied they seemed' (Haraway 2008: 24). Smuts realized that 'if she was really interested in these baboons, [she] had to enter into, not shun, a *responsive* relationship' (Haraway 2008: 25; emphasis added).[6] She 'soon learned that ignoring the proximity of another baboon is rarely a neutral act, something that should have been obvious to me from my experience among humans' (Smuts 2001: 297). Without words she got to know the baboons in what Haraway (2008: 25) calls 'the dance of relating', with all the connotations of embodied communication that a dance represents; dance as a material-semiotic reciprocal encounter and communication. Smuts and Haraway even go as far to suggest that there is more 'truth' in these closely interacting bodies than in speech, in which humans easily lie. It is a truth that is almost synonymous with honesty and authenticity. Following Smuts, responsiveness is all about the eye(s): 'The truth or honesty of non-linguistic embodied communication depends on looking back' (2001: 27). How relevant is such a fieldwork approach to the context of researching human animals? For inspiration to (begin to) answer this question we now turn our attention to disability studies (cf. Donaldson and Kymlicka 2011).

A trans-species research methodology?

In disability studies, we have come across (at least) two approaches to the development of research methodologies without words. The one is where a so-called disabled person, in this case Professor Temple Grandin, who is a self-declared autistic savant, argues, together with her co-author Catherine Johnson, that her autism gives her a unique ability 'to translate "animal talk" into English. I can tell people why their animals are doing the things they do' (Grandin and Johnson 2005: 7). She adds to this that 'what I was missing in social understanding [i.e., understanding humans] I could make up for in understanding animals' (ibid.). Grandin uses the word 'social' here as if it is only applicable to humans and could not also include non-human animals. But she goes even further than reducing 'the social' to humans only, by stating that '[a]nimals are like autistic savants. In fact, I'd go so far as to say that animals might actually *be* autistic savants [...] a difference in the brain autistic people share with animals' (Grandin and Johnson 2005: 8; emphasis original). Grandin has been severely criticized, especially for her statement that animals can basically be regarded as impaired humans (Vallortigara *et al.* 2008). One critic goes as far as to suggest that 'her ideas are way too problematic to even be quoted' as 'she seems oblivious to the amount of ideology she seems to be hiding under "biology" or just plainly supporting her weird take on disability and the animal mind' (Piskorski 2010: 1). Whatever you may think of this – and we do not share Grandin's general line of argument – we can agree with her and her co-author's hope that her book 'will help regular people be a *little less verbal* and a little more visual' (Grandin and Johnson 2005: 26; emphasis added). At least they recognize that in this field we are in need of research methodologies without words and in her book she explains how she does try to interpret the (social) behaviour of, in particular, farm animals.

When writing this chapter we realized we could not write about disability studies in relation to developing a trans-species research methodology, without mentioning Grandin's work. At the same time we want to explicitly state that we share the critiques that were quoted above that also strongly relate to the arguments that Bourke (2011) is making about the social constructedness of division among humans and between human and non-human animals. She makes abundantly clear that these distinctions are not at all about 'biological fact',[7] but about social, economic, and political power play and configurations. Grandin and Johnson (2005) explicitly pride themselves that, because of her particular insights into the animal mind, she has 'improved' the processes of slaughter and death of animals in the meat industry, but she has also been criticized by people who object to the meat industry, aptly summarized by Marc Bekoff (2010: 2):

> If I were a factory-farmed animal I would not look to Grandin to make my life better because the odds are that even if my life was slightly better it still

would be a horrible one at best, marked by interminable fear, terror, and anxiety (see also Bekoff 2013: 248–51).

Another inspirational example from disability studies is scientist David Goode (1995), who tries to understand the life of children born deaf and blind, who have not developed any formal language. Goode devotes a whole book to the methodological problems of how to research these children, and also 'how to use formal language to tell the story of persons themselves without formal language' (Goode 1995: 1). So, it is, first, about how to do the actual research (without words), how to reach a level of 'intersubjectivity' (Goode 1995: 8) with the children and, second, how to translate this intersubjectivity without words, into formal (scientific) language. Our impression and suggestion is that if we substitute 'children' for 'non-human animals', we are, once again, in the process of coming close to a trans-species research methodology without words, eclectically based on (interpreting and blending) 'embodied communication', 'bodily intersubjectivity', 'encounter', 'responsiveness', 'becoming animal', and 'becoming with'.

Goode's research on deaf, blind, and mentally impaired children and their communication with their parents relied primarily on 'bodily and gestural nonlanguage indexical expressions' (1995: 100). What seems to suggest that this could also work between human and non-human animals is the fact that Goode (ibid.) refers to in the following passage:

[t]he interpretation of these bodily indexical expressions was highly asymmetrical (those produced by the children were interpreted by us through language, and those produced by us were interpreted by children without language), but they were very clearly sufficient to permit communication, for all practical purposes, between us.

These ideas about indexical expressions date back to the famous ethnomethodologist Harold Garfinkel and his co-author Harvey Sacks. Key to indexical expressions is their contextual understanding: only when aware of the context in which the bodily expression is used, is one able to interpret its meaning, purpose, and direction. According to Goode (1995), contexts consist of the (clock) time, season, the type of occasion, biographical details, particular place and intentions, common, accumulated, gained or special contextual knowledge that can give the expression its obvious, but also more implicit, meanings. While indexical expressions were in, the first instance, developed to understand language and (therefore human) behaviour, they are also applied to 'non-oral communication' (and therefore also applicable to non-human animals) by Cicourel (Goode 1995: 101). Indexical expressions are contextual as much as they are bodily. In other words, it is about the 'body's expressivity', an approach to understanding human and non-human animal behaviour that has been preached and practiced ever since Darwin's 1872 publication of *The Expression of the Emotions in Man and Animals* (Darwin 2009). With Maurice Merleau-Ponty as its chief intellectual inspiration, 'the modern conception of bodily expressivity emphasizes its embeddedness in the world, and its meaning as emerging from specific involvement in that world, that is, *as indexical expressions*' (Goode 1995: 103; emphasis added). The particular communication (between species) involved in indexical and bodily expressions is that it is intentional (and therefore conscious); with the motility of the body, messages are constantly and mutually put across (cf. a 'responsive encounter'). In its embeddedness and contextuality it is always about communicating, expressing, and interpreting emotions. A beautiful and concrete description from Goode's book (1995: 111) is the following description of a morning greeting ritual between David Goode and Christina, one of the two deaf and blind children:

> I would greet Christina by placing her hand on my face. She would then gesture for me to pick her up. I would pick her up, and Chris would lock her legs around my waist and vigorously bounce up and down, indicating to me that she wanted me forcefully to throw or lift her up and down with my arms. I would do that often until my arms tired, which Chris could sense. She often would put her right (good) ear to my mouth, telling me she wanted me to talk or sing onto her ear while I bounced her up and down, and so forth. In this somewhat routinized but lively greeting, a kind of conservation with our bodies occurred, a communication with bodily signs that resembled in some ways a conversation with language.

An interesting sideline for the purpose of this particular chapter, in which we try to develop a trans-species research methodology without words, is Goode's (1995: 23) explicit statement that he got (part of) the inspiration for his research from the journals of Jean Itard, 'the French educator who took up the habilitation of the Wild Boy, Victoire of Aveyron'. Victoire was a so-called 'wolf-boy', 'a child raised in the woods' (Goode 1995: 219). It would be interesting to know if Goode would consider the research methodologies he developed and perfected during his research of deaf, blind, and mentally impaired children, to also have potential for researching non-human animals. There seem to be some hints that this might be the case, as we recognize some of his suggestions in the research that for instance Barbara Smuts and Dian Fossey conducted among baboons and gorillas: 'I stopped trying to teach her, and began to let her teach me. I decided to *mimic her actions* in order to gain more direct access to what such activities were providing her' (Goode 1995: 33; emphasis added). Both Smuts and Fossey mimicked the behaviour of the baboons and gorillas for the same reasons. In other words, they tried to learn from them, and in the process became responsive. In ethnographic terms this could be referred to as a form of participant observation. Participant observation as a qualitative method is already difficult for any ethnographer researching human configurations (cf. Ybema *et al.* 2009), but comes with a whole set of extra challenges when we try to apply it across species. When trying to bring this message home in the context of greeting rituals, we are reminded of Shaun Ellis's (Ellis and Junor 2009: 99–106) description of his greeting rituals with wolves, which involved a great deal of mutual licking and touching tongues. Many people do not mind being licked by a dog (every now and then), but to touch tongues would probably be loathed by most of us. It requires a type of intimacy that we (perhaps) dare or are brought to share with humans, but which many people would probably not like to share with non-human animals, e.g., eating the same food, participating in daily rituals, or other even more personal intimacies. The same applies to another story that Ellis relates, telling of his time among a pack of wild wolves where he occupied the most subordinate and powerless position available. In order to show submissiveness, he had to lie on his back and keep completely still, while the huge fangs of the more powerful wolf in the pack were around his throat. This is a form of trans-species participant observation that indeed teaches us a lot about wolves, but definitely comes with its own personal challenges.

What makes this example a fitting prelude to the conclusion of this chapter is the opening story of Ellis's book where many aspects of trans-species communication, and the various elements and topics we touched upon in this chapter, seem to come together (and we quote at length here, to let the reader appreciate its rhetorical power):

> I was helping out at a wildlife center in Hertfordshire, one of the home counties, just north of London. A man appeared outside the wolf enclosure one day, pushing a child in an old-fashioned wheelchair that looked almost Victorian, with a large rectangular tray on the front of it. I was immediately struck by how out of place it looked. He told me that he and his son, who may have been thirteen or fourteen and who, I could see at a glance, was severely disabled, had driven all the way from Scotland, a distance of around five hundred miles. [...] I could immediately see there would be a problem and asked the father, as tactfully as I could, whether the child would be able to indicate when he no longer wanted to be near the wolf, explaining how important this was. 'He won't be able to', said the man, bluntly. 'He has never spoken, and never reacted in any way to anything. And he has never expressed an emotion in his life.' Questioning my sanity, I went into the enclosure and came out carrying Zarnesti. He was then about three months old, the size of a spaniel and a wriggling, struggling bundle of energy. It was all I could do to hold him; he was almost flying out of my arms as I put him down onto the tray on this old-fashioned wheelchair in front of the boy. I had the pup in a viselike grip, but something miraculous happened. The moment Zarnesti saw the child he became still. He looked into the boy's eyes and they stared at each other. Then the pup settled down with his back legs tucked under him and his front legs stretched out in front. I took one hand off him and I realized very quickly that I could take the other hand away, too. After a few moments, still looking into his eyes, the cub reached forward and started to lick the boy's face. I lunged to intercept him, terrified that Zarnesti would nip the boy's mouth with his needle-sharp teeth, which is what cubs do to adult wolves when they want them to regurgitate food. But Zarnesti didn't nip; he just licked, very gently. The scene was electrifying. As I looked at the boy I saw one single tear welling up in his right eye, then trickle down his

cheek. Guessing this had never happened before, I turned to his father. This big, strong, capable Scotsman was standing, watching what was unfolding in front of him, with tears streaming down his face. In a matter of seconds, the wolf cub had gotten through to this boy in a way that no human had managed to do in fourteen years (Ellis and Junor 2009: xi–xiv).[8]

To return to Goode (1995) once more, what could be considered a beautiful side-effect of this approach to understanding indexical expressions in disabled children or humans in general, is the observation by Goode (1995: 113–4; emphasis added): 'when one understood the concrete projects [i.e., contextualized the bodily expressions], the membership of those performing them, their place and their purpose, and so forth, the actions of these children *lost* their pathological quality'. If we apply this to understanding trans-species communication between human and non-human animals, a trans-species research methodology without words could well breach the hierarchical, deprecating, and asymmetrical relations that our man-made and human-exceptionalism-inspired distinctions always bring along with them when studying non- or not-so-human others (including disabled or racially different humans, cf. Bourke 2011; DeKoven and Lundblad 2012).

Tentative conclusions

Anthropology is a discipline that, because of its history and development, should be ideally positioned to further the development of trans-species research and thinking. Non-human animals have always played a role in anthropological research and fieldwork, but nevertheless, as a discipline, anthropology has become, both theoretically and methodologically, rather anthropocentric, if not outright speciesist, in its orientation. This is a feature of the discipline that should be seen in its time and age where social sciences in general were rather anthropocentric. But anthropology has also been shown to be adaptable, and so it will be in terms of including the non-human animal in its research. Although considerable attention has been devoted to the theoretical implication of the 'animal turn', research methodology has so far received far less attention.

This chapter is part of our ongoing attempts and processes to come to grips with developing trans-species research methodologies without words in order to be able to include both human and non-human animals (and their relationships) in a single research approach. This is necessarily a research methodology without words, as we cannot interview or otherwise rely on them in our exchanges with and research on non-human animals. Our inspiration comes from a combination of conceptual and rather philosophically inspired discussions on empathy and research in the field of disability studies. It is an approach in which we basically follow Calarco's (2008) conclusion and suggestion that to research and understand human–animal relations, we simply have to 'let go' of the human–animal distinction. In the wake of this revolutionary call we can start to try and think more systematically on how to develop trans-species research methodologies without words that might help to start to integrate Calarco's (ibid.) 'zoographies' with Donaldson and Kymlicka's (2011) political ideas for a 'zoopolis', where human and (several categories of) non-human animals share a common society and citizenship.

Notes

1 In his foreword to David Goode's (1995) book, Irving Kenneth Zola (1995: ix) states that 'in theories of the formal organization of society and culture, we have overstressed the essentialness of spoken speech'.

2 A part of this section has been published previously in the essay by Harry Wels, 'Whispering empathy: transdisciplinary reflections on research methodology' in B. Musschenga and A. Van Harskamp, eds. *What makes us moral: on the capacities and conditions for being moral.* Dordrecht, Netherlands: Springer, 151–65. We wish to thank Springer Publishers for the opportunity to reuse the section.

3 Liebenberg (1990) reports similar observations about the San in southern Africa. 'In order to understand animals, the [San] trackers must identify themselves with an animal' (Liebenberg 1990: 88). Therefore, '[t]he [San] knowledge of animal behavior essentially has an anthropomorphic nature', and '[a]lthough their knowledge is at variance with that of European ethologists, it has withstood the vigorous empirical testing imposed by its use'. Therefore, '[a]nthropomorphism may well have its origins in the way trackers must identify themselves with an animal' (Liebenberg 1990: 83).

4 It is important to keep in mind this concept of 'collective of contexts', for later on in the chapter we will discuss the concept of 'indexical expressions'. Mary Midgley (1995: 334–5) also emphasizes the importance of contexts in her work: 'If one asks, "is this genuinely nest-building or displacement activity?" one is asking about contexts – what else has it been doing, what will it go on to do. And the answer will involve all kinds of details about its behavior and the other contexts in which these same actions have previously appeared'. She applies it to interpreting both human and nonhuman animal behaviour.

5 Levinas, however, did not apply his idea of the encounter to non-human animals.

6 In a non-language environment this is usually captured by referring to 'empathetic understanding'. This type of understanding is often related to contexts of care (cf. Haraway 2008: 82–93). Empathy and care often come and go together in popular books on relations between human and non-human animals, in particular when referring to relations between humans and so-called 'wild animals'. Especially on the African continent, that is often (touristically) caricatured and stereotyped for its wildlife, there is a whole genre of books that narrate cases of, in particular, white people empathizing with wildlife in the contexts of care. For a recent example see Sheldrick (2012), whose dust cover reads: 'Daphne Sheldrick is the first person to have successfully hand-reared newborn elephants. Her deep empathy and understanding […] ha[s] saved countless elephants'. See, in addition, the very popular series of books on Elsa the Lioness by Joy Adamson, also combining empathy and care.

7 Piskorski (2010) writes that Grandin is constantly 'quoting of biological fact as if she were quoting the Holy Writ'.

8 In South Africa a recent study shows that using elephants to interact with handicapped children also offers huge potential (Dronkers 2012). In the Netherlands there is a party-like political movement that devotes its highest priority 'to animal welfare and the respectful treatment of animals' (Party for the Animals 2015). Also an extraordinary professorship has been awarded to Professor Marie-José Enders-Slegers at the Open University, the first chair in Europe devoted to the emerging field of 'anthrozoology' focusing on human–animal interaction in contexts of care (see Open Universiteit 2013).

References

Arluke, A. and Sanders, C., 1996. *Regarding animals*. Pennsylvania, US: Temple University Press.

Asquith, P.J., ed., 2002. *A Japanese view of nature: the world of living things by Kinji Imanishi*. London: Routledge Curzon.

Bamberg, M., 2007. *Narrative – state of the art*. Amsterdam: John Benjamins.

Bekoff, M., 2007. *The emotional lives of animals: a leading scientist explores animal joy, sorrow, and empathy – and why they matter*. Novato, US: New World Library.

Bekoff, M., 2010. Going to slaughter: should animals hope to meet Temple Grand[in] [Online]. Available from: https://www.psychologytoday.com/blog/animal-emotions/201002/going-slaughter-should-animals-hope-meet-temple-grand [Accessed 26 February 2015].

Bekoff, M., 2013. *Why dogs hump and bees get depressed: the fascinating science of animal intelligence, emotions, friendship, and conservation*. Novato, US: New World Library.

Bourke, J., 2011. *What it means to be human: reflections from 1791 to the present*. London: Virago.

Buller, H., 2014. Animal Geographies I. *Progress in Human Geography*, 38 (2), 308–18.

Calarco, M., 2008. *Zoographies: the question of the animal from Heidegger to Derrida*. New York: Columbia University Press.

Darwin, C., 2009 [1872]. *The expression of the emotions in man and animals*. London: Penguin.

DeKoven, M. and Lundblad, M., eds, 2012. *Species matter: humane advocacy and cultural theory*. New York: Columbia University Press.

Deleuze, G. and Guattari, F., 1980. *A thousand plateaus: capitalism and schizophrenia*. Minnesota, US: University of Minnesota Press.

Derrida, J., 2008. *The animal that therefore I am*. New York: Fordham University Press.

De Waal, F., 2001. *The ape and the sushi master: reflections of a primatologist*. New York: Basic Books.

De Waal, F., 2008. Putting the altruism back into altruism: the evolution of empathy. *Annual Review of Psychology*, 59, 279–300.

De Waal, F., 2009. *The age of empathy: nature's lessons for a kinder society*. New York: Harmony Books.

Donaldson, S. and Kymlicka, W., 2011. *Zoopolis: a political theory of animal rights*. Oxford, UK: Oxford University Press.

Dronkers, M., 2012. Care farming in South Africa: possible or not? A case study at Adventures with Elephants. Unpublished. Master's thesis. Amsterdam, Netherlands: Vrije Universiteit Amsterdam.

Ellis, S. and Junor, P., 2009. *The man who lives with wolves*. New York: Harmony Books.

Evans-Pritchard, E.E., 1956. *Nuer religion*. New York: Oxford University Press.

Fox Keller, E., 1983. *A feeling for the organism: the life and work of Barbara McClintock*. New York: W.H. Freeman and Company.

Geertz, C., 1973. *The interpretation of cultures*. New York: Basic Books.

Goode, D., 1995. *A world without words: the social construction of children born deaf and blind*. Philadelphia, US: Temple University Press.

Grandin, T. and Johnson, C., 2005. *Animals in translation: using the mysteries of autism to decode animal behavior*. New York: Scribner.

Haraway, D., 2008. *When species meet*. Minneapolis, US: University of Minnesota Press.

Hutchinson, S.E., 1996. *Nuer dilemmas: coping with money, war, and the state*. Berkeley, US: University of California Press.

Janssens, M. and Steyaert, C., 2001. *Meerstemmigheid: organiseren met verschil.* Leuven, Belgium: Universitaire Pers Leuven.

King, B.J., 2013. *How animals grieve.* Chicago, US: University of Chicago Press.

Kirksey, S.E. and Helmreich, S., 2010. The emergence of multi-species ethnography. *Cultural Anthropology*, 25 (4), 545–76.

Kruuk, H., 2003. *Niko's nature: a life of Niko Tinbergen and his science of animal behavior.* Oxford, UK: Oxford University Press.

Lestel, D., Brunois, F., and Gaunet, F., 2006. Etho-ethnology and ethno-ethology. *Social Science Information*, 45 (2), 155–76.

Liebenberg, L., 1990. *The art of tracking: the origin of science.* Claremont, South Africa: David Philip.

Marshall Thomas, E., 2009 [1991]. Foreword. *In*: S. Montgomery. *Walking with the great apes: Jane Goodall, Dian Fossey, Biruté Galdikas.* White River Junction, US: Chelsea Green Publishing, xi–xvi.

Midgley, M., 1995 [1975]. *Beast and man.* London: Routledge.

Montgomery, S., 2009 [1991]. *Walking with the great apes: Jane Goodall, Dian Fossey, Biruté Galdikas.* White River Junction, US: Chelsea Green Publishing.

Open Universiteit, 2013. Persberichten: eerste Europese leerstoel in Nederland over bijdrage dier aan zorg en welzijn mens [Online]. Available from: www.ou.nl/web/persberichten/home/-/asset_publisher/6fwK/content/id/2039813 [Accessed 27 February 2015].

Party for the Animals, 2015. Home [Online]. Available from: www.partyfortheanimals.nl/ [Accessed 26 February 2015].

Piskorski, R., 2010. Temple Grandin's ableism [Online]. Available from: http://posthumanities.blogspot.com/2010/02/temple-gradins-ableism.html [Accessed 26 February 2015].

Ryan, M.J., 2010. An improbable path. *In*: L. Drickamer and D. Dewsbury, eds. *Leaders in animal behavior: the second generation.* Cambridge, UK: Cambridge University Press, 465–95.

Serpell, J., 1986. *In the company of animals: a study of human–animal relationships.* New York: Cambridge University Press.

Sheldrick, D., 2012. *Love, life and elephants: an African love song.* London: Viking.

Smuts, B., 2001. Encounters with animal minds. *In*: E.T. Thompson, ed. *Between ourselves: second-person issues in the study of consciousness.* Thorverton, UK: Imprint Academic, 293–309.

Takasaki, H., 2002. Foreword. *In*: P.J. Asquith, ed. *A Japanese view of nature: the world of living things by Kinji Imanishi.* London: Routledge Curzon, xi–xii.

Ten Bos, R., 2008. *Het geniale dier: een andere antropologie.* Amsterdam: Boom.

Vallortigara, G., Snyder, A., Kaplan, G., Bateson, P., Clayton, N.S., and Rogers, L.J., 2008. Are animals autistic savants? *PLoS Biology* [Online], 6 (2). Available from: doi:10.1371/journal.pbio.0060042 [Accessed 26 February 2015].

Wels, H., 2013. Whispering empathy: transdisciplinary reflections on research methodology. *In*: B. Musschenga and A. Van Harskamp, eds. *What makes us moral: on the capacities and conditions for being moral.* Dordrecht, Netherlands: Springer, 151–65.

Woodward, W., 2008. *The animal gaze: animal subjectivities in southern African narratives.* Johannesburg, South Africa: Wits University Press.

Ybema, S., Yanow, D., Wels, H., and Kamsteeg, F., eds, 2009. *Organizational ethnography: studying the complexity of everyday life.* Thousand Oaks, US: Sage.

Zola, I.K., 1995. Foreword. *In*: D. Goode. *A world without words: the social construction of children born deaf and blind.* Philadelphia, US: Temple University Press, ix–xii.

Index

Printed in the United States
by Baker & Taylor Publisher Services